STUDENT SOLUTIONS MANUAL

Jeffrey Witmer
Oberlin College

Contributions by
Carol Adjemian
Pepperdine University

Statistics for the
Life Sciences
THIRD EDITION

SAMUELS • WITMER

PEARSON
Prentice Hall

Upper Saddle River, NJ 07458

Editor-in-Chief: Sally Yagan
Senior Acquisitions Editor: Petra Recter
Supplement Editor: Joanne Wendelken
Executive Managing Editor: Kathleen Schiaparelli
Assistant Managing Editor: Karen Bosch
Production Editor: Jenelle J. Woodrup
Supplement Cover Manager: Paul Gourhan
Supplement Cover Designer: Christopher Kossa
Manufacturing Buyer: Ilene Kahn
Manufacturing Manager: Alexis Heydt-Long

© 2006 Pearson Education, Inc.
Pearson Prentice Hall
Pearson Education, Inc.
Upper Saddle River, NJ 07458

Printed in the United States of America

10 9 8 7 6 5 4 3 2 1

ISBN 0-13-041317-8

Pearson Education Ltd., *London*
Pearson Education Australia Pty. Ltd., *Sydney*
Pearson Education Singapore, Pte. Ltd.
Pearson Education North Asia Ltd., *Hong Kong*
Pearson Education Canada, Inc., *Toronto*
Pearson Educación de Mexico, S.A. de C.V.
Pearson Education—Japan, *Tokyo*
Pearson Education Malaysia, Pte. Ltd.

CONTENTS

CHAPTER 2

Description of Samples and Populations

2.2 (a) i) height and weight
ii) continuous variables
iii) a child
iv) 37

(b) i) blood type and cholesterol level
ii) blood type is categorical, cholesterol level is continuous
iii) a person
iv) 129

2.4 (a) There is no single correct answer. The smallest observation is 5.4 mm and the largest is 6.7 mm; thus, the range to be spanned is 6.7 - 5.4 = 1.3. We choose to use 7 classes with a class width of .2 (because (7)(.2) = 1.4 is wide enough to span the data.) We choose the first class limit as 5.4 (because it is a multiple of .2, *not* because it is the smallest observation) and obtain the following table and histogram.

Molar width	Frequency (no. specimens)
5.4-5.5	1
5.6-5.7	5
5.8-5.9	7
6.0-6.1	12
6.2-6.3	8
6.4-6.5	2
6.6-6.7	1
Total	36

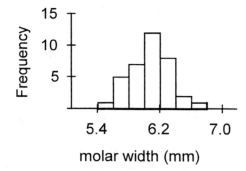

A choice of 5 classes with a class width of .3 would also be reasonable.

(b) The distribution is fairly symmetric.

2.10 There is no single correct answer. One possibility is

Glucose (%)	Frequency (no. of dogs)
70-74	3
75-79	5
80-84	10
85-89	5
90-94	2
95-99	2
100-104	1
105-109	1
110-114	0
115-119	1
120-124	0
125-129	0
130-134	1
Total	31

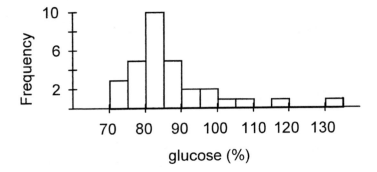

2.11

```
 7 | 8 4 0 6 0 9 5 5
 8 | 1 8 5 4 1 4 2 6 9 9 0 2 1 4 2
 9 | 3 3 9 6
10 | 2 6
11 | 5
12 |
13 | 1
```
Key 7|8 = 78%

2.16 $\overline{y} = \Sigma y_i/n = \dfrac{6.3 + 5.9 + 7.0 + 6.9 + 5.9}{5} = 6.40$ nmol/gm. The median is the 3rd largest value (i.e., the third observation in the *ordered* array of 5.9 5.9 6.3 6.9 7.0), so the median is 6.3 nmol/gm.

2.18 $\bar{y} = \Sigma y_i/n = \dfrac{366+327+274+292+274+230}{6} = 293.8$ mg/dl. The median is the average of the 3rd and 4th largest values (i.e., the third and fourth observations in the *ordered* array of 230 274 274 292 327 366), so the median is $\dfrac{274 + 292}{2} = 283$ mg/dl.

2.24 The median is the average of the 18th and 19th largest values. There are 18 values less than or equal to 10 and 18 values that are greater than or equal to 11. Thus, the median is $\dfrac{10 + 11}{2} = 10.5$ piglets.

2.26 The distribution is fairly symmetric so the mean and median are roughly equal. It appears that half of the distribution is below 40 and half is above 40. Thus, mean $\approx$ median $\approx$ 40.

2.30 (a) Putting the data in order, we have

$$13\ \ 13\ \ 14\ \ 14\ \ 15\ \ 15\ \ 16\ \ 20\ \ 21\ \ 26$$

The median is the average of observations 5 and 6 in the ordered list. Thus, the median is $\dfrac{15 + 15}{2} = 15$. The lower half of the distribution is

$$13\ \ 13\ \ 14\ \ 14\ \ 15$$

The median of this list is the 3rd largest value, which is 14. Thus, the first quartile of the distribution is $Q_1 = 14$. Likewise, the upper half of the distribution is

$$15\ \ 16\ \ 20\ \ 21\ \ 26$$

The median of this list is the 3rd largest value, which is 20. Thus, the third quartile of the distribution is $Q_3 = 20$.

(b) IQR = Q_3 - Q_1 = 20 - 14 = 6

(c) To be an outlier at the upper end of the distribution, an observation would have to be larger than Q_3 + 1.5(IQR) = 20 + 1.5(6) = 20 + 9 = 29, which is the upper fence.

2.31 (a) The median is the average of the 9th and 10th largest observations. The ordered list of the data is

$$4.1\ \ 5.2\ \ 6.8\ \ 7.3\ \ 7.4\ \ 7.8\ \ 7.8\ \ 8.4\ \ 8.7\ \ 9.7\ \ 9.9\ \ 10.6\ \ 10.7\ \ 11.9\ \ 12.7\ \ 14.2\ \ 14.5\ \ 18.8$$

Thus, the median is $\dfrac{8.7 + 9.7}{2} = 9.2$.

To find Q_1 we consider only the lower half of the data set:

$$4.1\ \ 5.2\ \ 6.8\ \ 7.3\ \ 7.4\ \ 7.8\ \ 7.8\ \ 8.4\ \ 8.7\ \ 9.7$$

Q_1 is the median of this half (i.e., the 5th largest value), which is 7.4.
To find Q_3 we consider only the upper half of the data set:

$$9.7\ \ 9.9\ \ 10.6\ \ 10.7\ \ 11.9\ \ 12.7\ \ 14.2\ \ 14.5\ \ 18.8.$$

Q_3 is the median of this half (i.e., the 5th largest value in this list), which is 11.9.

(b) IQR = Q_3 - Q_1 = 11.9 - 7.4 = 4.5

(c)

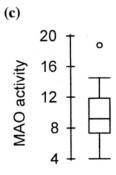

2.40 (a) $\bar{y} = 15$, $\Sigma(y_i - \bar{y})^2 = 18$, $s = \sqrt{18/3} = 2.45$

   (b) $\bar{y} = 35$, $\Sigma(y_i - \bar{y})^2 = 44$, $s = \sqrt{44/4} = 3.32$

2.43 (a) $\bar{y} = 33.10$ lb; $s = 3.44$ lb

   (b) Coefficient of variation $= \dfrac{s}{\bar{y}} = \dfrac{3.44}{33.10} = .104$ or 10.4%

2.45 The data are -13, -29, -7, 2, -10, -43, 4, 15, -13, -30. $\bar{y} = -12.4$ mm Hg; $s = 17.6$ mm Hg.

2.53 Coefficient of variation $= \dfrac{s}{\bar{y}} = \dfrac{6.8}{166.3} = .04$ or 4%

2.56 $y' = (y - 7)*100$. Thus, the mean of y' is $(\bar{y} - 7)*100 = (7.373 - 7)*100 = 37.3$. The SD of y' is $s*100 = .129*100 = 12.9$.

2.65 (a)

```
 9 | 8 8 9
10 | 0 0 6 7 7 7 8
11 | 0 0 1 4 5 5 6 6 6 9
12 | 0 1 2 3 3 4
13 | 0
```

Key 9|8 = .098

2.73 (a) n = 119, so the median is the 60th largest observation. There are 32 observations less than or equal to 37 and 44 observations less than or equal to 38. Thus, the median is 38.

   (b) The first quartile is the 30th largest observation, which is 36. The third quartile is the 90th largest observation, which is 41.

   (d) The mean is 38.45 and the SD is 3.20. Thus, the interval $\bar{y} \pm s$ is 38.45 ± 3.20, which is 35.25 to 41.65. This interval includes 36, 27, 28, 29, 40, and 41. The number of flies with 36 to 41 bristles is $11 + 12 + 18 + 13 + 10 + 15 = 79$. Thus, the percentage of observations that fall within one standard deviation of the mean is $79/119*100\% = .664*100\% = 66.4\%$.

CHAPTER 3

Random Sampling, Probability, and the Binomial Distribution

3.5 (a) In the population, 51% of the fish have 21 vertebrae. Thus, $\Pr\{Y = 21\} = .51$.

   (b) In the population, the percentage of fish with 22 or fewer vertebrae is $3 + 51 + 40 = 94\%$. Thus, $\Pr\{Y \leq 22\} = .94$.

3.10 (a)

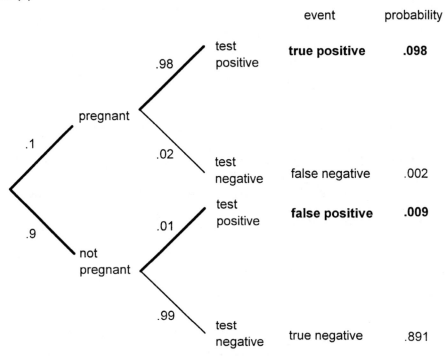

	event	probability
test positive	**true positive**	**.098**
test negative	false negative	.002
test positive	**false positive**	**.009**
test negative	true negative	.891

   There are two ways to test positive. A true positive happens with probability $(.1)(.98) = .098$. A false positive happens with probability $(.9)(.01) = .009$. Thus, $\Pr\{\text{test positive}\} = .098 + .009 = .107$.

   (b) Using the same reasoning as in part (a),
   $\Pr\{\text{test positive}\} = (.05)(.98) + (.95)(.01) = .049 + .095 = .0585$.

3.16 (a) $\Pr\{20 < Y < 30\} = .41 + .21 = .62$

   (b) $.41 + .21 + .03 = .65$

   (c) $.01 + .34 = .35$

3.31 On average, there are 105 males to every 100 females. Thus, $\Pr\{\text{male}\} = \frac{105}{205}$ and $\Pr\{\text{female}\} = \frac{100}{205}$. To use the binomial distribution, we arbitrarily identify "success" as "female."

(a) We have $n = 4$ and $p = \frac{100}{205}$. To find the probability of 2 males and 2 females, we set $j = 2$, so $n - j = 2$. The binomial formula gives $\Pr\{2 \text{ males and } 2 \text{ females}\} = {}_4C_2 \left(\frac{100}{205}\right)^2 \left(\frac{105}{205}\right)^2 = .3746$.

(b) To find the probability of 4 males, we set $j = 0$, so $n - j = 4$. The binomial formula gives $\Pr\{4 \text{ males}\} = {}_4C_0 \left(\frac{100}{205}\right)^0 \left(\frac{105}{205}\right)^4 = (1)(1)\left(\frac{105}{205}\right)^4 = .0688$.

(c) The condition that all four infants are the same sex can be satisfied two ways: All four could be male or all four could be female. The probability that all four are male has been computed in part (b) to be .0688. To find the probability that all four are females, we set $j = 4$, so $n - j = 0$. $\Pr\{4 \text{ females}\} = {}_4C_4 \left(\frac{100}{205}\right)^4 \left(\frac{105}{205}\right)^0 = (1)(1)\left(\frac{100}{205}\right)^4 = .0566$. Thus, we find that $\Pr\{\text{all four are the same sex}\} = .0566 + .0688 = .1254$.

3.34 $\Pr\{\text{high blood level}\} = \frac{1}{8} = p$. To apply the binomial formula, we arbitrarily identify "success" as "high blood lead." Then $n = 16$ and $p = \frac{1}{8}$.

(a) To find the probability that none has high blood lead, we set $j = 0$, so $n - j = 16$. The binomial formula gives $\Pr\{\text{none has high blood lead}\} = {}_{16}C_0 \left(\frac{1}{8}\right)^0 \left(\frac{7}{8}\right)^{16} = (1)(1)\left(\frac{7}{8}\right)^{16} = .1181$.

(b) To find the probability that one has high blood lead, we set $j = 1$, so $n - j = 15$. The binomial formula gives $\Pr\{\text{one has high blood lead}\} = {}_{16}C_1 \left(\frac{1}{8}\right)^1 \left(\frac{7}{8}\right)^{15} = (16)\left(\frac{1}{8}\right)\left(\frac{7}{8}\right)^{15} = .2699$.

(c) To find the probability that two have high blood lead, we set $j = 2$, so $n - j = 14$. The binomial formula gives $\Pr\{\text{two have high blood lead}\} = {}_{16}C_2 \left(\frac{1}{8}\right)^2 \left(\frac{7}{8}\right)^{14} = (120)\left(\frac{1}{8}\right)^2 \left(\frac{7}{8}\right)^{14} = .2891$.

(d) $\Pr\{\text{three or more have high blood lead}\} = 1 - \Pr\{2 \text{ or fewer have high blood lead}\}$
$= 1 - [.1181 + .2699 + .2891] = .3229$.

3.35 The first step is to determine the best-fitting value for $p = \Pr\{\text{boy}\}$. The total number of children in all the families is
$(6)(72,069) = 432,414$.

The number of boys is
$$(0)(1{,}096) + (1)(6{,}233) + ... + (6)(1{,}579) = 222{,}638.$$

Thus, the value of p that fits the data best is
$$p = \frac{222638}{432414} = .514872321$$

To compute the probabilities of various sex ratios, we apply the binomial formula with n = 6 and p = .514872321. Then we multiply each probability by 72,069 to obtain the expected frequency:

Number of boys (j)	Expected frequency		
0	$(72{,}069)(1)(1 - p)^6$	=	939.5
1	$(72{,}069)(6)(p)(1 - p)^5$	=	5,982.5
2	$(72{,}069)(15)(p^2)(1 - p)^4$	=	15,873.1
3	$(72{,}069)(20)(p^3)(1 - p)^3$	=	22,461.8
4	$(72{,}069)(15)(p^4)(1 - p)^2$	=	17,879.3
5	$(72{,}069)(6)(p^5)(1 - p)^1$	=	7,590.2
6	$(72{,}069)(1)(p^6)$	=	1,342.6

The following table compares the observed and expected frequencies:

Number of boys	Number of girls	Observed frequency	Expected frequency	Sign of (Obs - Exp)
0	6	1,096	939.5	+
1	5	6,233	5,982.5	+
2	4	15,700	15,873.1	-
3	3	22,221	22,461.8	-
4	2	17,332	17,879.3	-
5	1	7,908	7,590.2	+
6	0	1,579	1,342.6	+
		72,069	72,069.0	

We note that there is reasonable agreement between the observed and expected frequencies. However, the observed frequencies exceed the expected frequencies for the preponderantly unisex siblingships (those with 0, 1, 5, or 6 boys), whereas the observed frequencies are less than the expected frequencies for the more balanced siblingships (2, 3, or 4 boys). This pattern is similar to that seen in Example 3.35.

3.40 The probability that a square has no centipedes is the relative frequency of squares with zero centipedes. Thus, Pr{no centipedes} = .45.

To apply the binomial formula, we identify "success" as "no centipedes." Then n = 5 and p = .45. To find the probability that three squares have centipedes and two do not, we set j = 2, so n - j = 3.

$$Pr\{3 \text{ with, } 2 \text{ without}\} = {}_5C_2(.45^2)(.55^3) = 10(.45^2)(.55^3) = .3369.$$

3.45 The probability that an *innocent* subject will have a higher electrodermal response on any critical word (compared to the three controls for that critical word) is 25%. Thus, Pr{failure on a critical word} = .25.

Let us identify "success" as "success on a critical word." Thus, p = .75 and 1 - p = .25. We then use the binomial formula with n = 6 and p = .75.

To be labeled "guilty" a subject must fail 4 or more critical words out of 6. We find the probability of 4, 5, or 6 failures by setting j = 2, 1, or 0, as follows:

$$\Pr\{2 \text{ successes, 4 failures}\} = {}_6C_2 p^2 (1-p)^4 = (15)(.75^2)(.25^4) = .0330$$
$$\Pr\{1 \text{ success, 5 failures}\} = {}_6C_1 p^1 (1-p)^5 = (6)(.75^1)(.25^5) = .0044$$
$$\Pr\{0 \text{ successes, 6 failures}\} = {}_6C_0 p^0 (1-p)^6 = (1)(1)(.25^6) = .0002$$

To find the probability that an innocent subject will be labeled "guilty," we add these three values:

Pr{4 or more failures} = .0330 + .0044 + .0002 = .0376.

3.47 To use the binomial distribution here, let us identify "success" as "blood pressure > 140." Then, p = Pr{blood pressure > 140} = .25 + .09 + .04 = .38, which is the area under the density curve beyond 140 mm Hg.

The number of trials is n = 4. To find the probability that all four men have blood pressure higher than 140 mm Hg, we set j = 4, so n - j = 0.

$$\Pr\{\text{all four have blood pressure} > 140\} = {}_4C_4 p^4 (1 - p)^0 = (1)(.38^4)(1) = .0209.$$

(b) $\Pr\{\text{three have blood pressure} > 140\} = {}_4C_3 p^3 (1 - p)^1 = (4)(.38^3)(.62^1) = .1361.$

CHAPTER 4

The Normal Distribution

4.3 $\mu = 1400$; $\sigma = 100$.

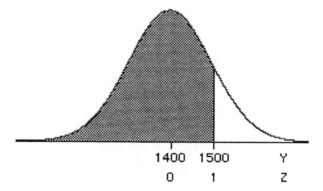

(a) For y = 1500,

$$z = \frac{y - \mu}{\sigma} = \frac{1500 - 1400}{100} = 1.00.$$

From Table 3, the area is .8413 or 84.13%.

(b) For y = 1325,

$$z = \frac{y - \mu}{\sigma} = \frac{1325 - 1400}{100} = -.75.$$

From Table 3, the area below 1325 is .2266.
From part (a), the area below 1500 is .8413.
Thus, the percentage with
$$1325 \leq Y \leq 1500 \text{ is}$$
.8413 - .2266 = .6147 or 61.47%.

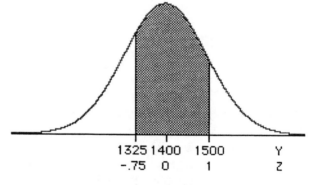

(c) For y = 1325,

$$z = \frac{y - \mu}{\sigma} = \frac{1325 - 1400}{100} = -.75.$$

From Table 3, the area below 1325 is .2266.
Thus, the percentage with $Y \geq 1325$ is
1 - .2266 = .7734 or 77.34%.

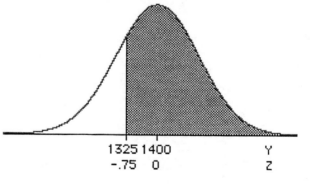

10

(d) For y = 1475,

$$z = \frac{y - \mu}{\sigma} = \frac{1475 - 1400}{100} = .75.$$

From Table 3, the area below 1475 is .7734.
Thus, the percentage with Y ≥ 1475 is
1 - .7734 = .2266 or 22.66%.

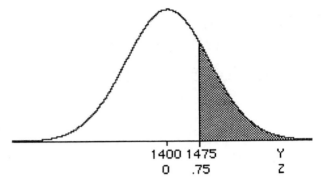

(e) For y = 1600,

$$z = \frac{y - \mu}{\sigma} = \frac{1600 - 1400}{100} = 2.00.$$

From Table 3, the area below 1600 is .9772.
In part (d) we found that the area below
    1475 is .7734.
Thus, the percentage with
    1475 ≤ Y ≤ 1600 is
.9772 - .7734 = .2038 or 20.38%.

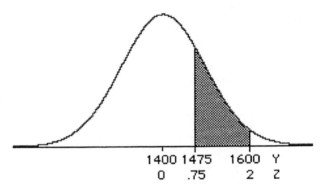

(f) For y = 1200,

$$z = \frac{y - \mu}{\sigma} = \frac{1200 - 1400}{100} = -2.00.$$

From Table 3, the area below 1200 is .0228.
In part (c) we found that the area below
    1325 is .2266.
Thus, the percentage with
    1200 ≤ Y ≤ 1325 is
.2266 - .0028 = .2038 or 20.38%.

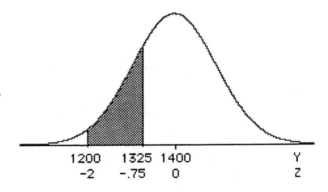

4.4 μ = 1400; σ = 100.

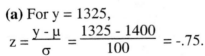

(a) For y = 1325,

$$z = \frac{y - \mu}{\sigma} = \frac{1325 - 1400}{100} = -.75.$$

From Table 3, the area is .2266 or 22.66%.

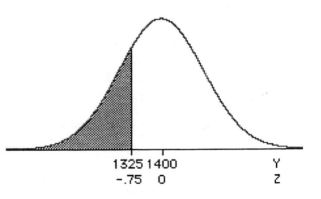

(b) This is the same as part (e) of Exercise 4.3.

For y = 1600,

$$z = \frac{y - \mu}{\sigma} = \frac{1600 - 1400}{100} = 2.00.$$

From Table 3, the area below 1600 is .9772.

For y = 1475,

$$z = \frac{y - \mu}{\sigma} = \frac{1475 - 1400}{100} = .75.$$

From Table 3, the area below 1475 is .7734.

Thus, the percentage with
 $1475 \le Y \le 1600$ is
.9772 - .7734 = .2038 or 20.38%.

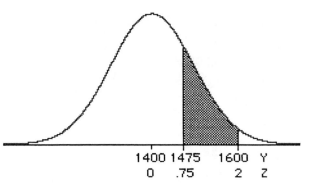

4.8 $\mu = 88$; $\sigma = 7$.

(a) The 65th percentile is the value that is larger than 65% of the observations. Thus, the area under the curve below this value is .65.

In Table 3, the area closest to .65 is .6517, which corresponds to z = .39.

Thus, the 65th percentile y* must satisfy the equation

$$z = \frac{y^* - \mu}{\sigma} \quad \text{or} \quad .39 = \frac{y^* - 88}{7}.$$

The solution of this equation is
 $y^* = (7)(.39) + 88 = 90.7.$

Thus, the 65th percentile is 90.7 lb.

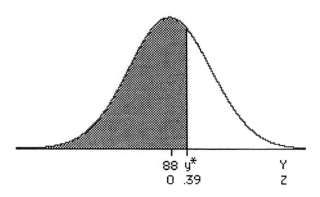

(b) The 35th percentile is the value that is larger than 35% of the observations. Thus, the area under the curve below this value is .35.

In Table 3, the area closest to .35 is .3483, which corresponds to z = -.39.

(Note that this is the negative of the value found in part (a).)

Thus, the 35th percentile y* must satisfy the equation

$$z = \frac{y^* - \mu}{\sigma} \quad \text{or} \quad -.39 = \frac{y^* - 88}{7}.$$

The solution of this equation is
 $y^* = (7)(-.39) + 88 = 85.3.$

Thus, the 35th percentile is 85.3 lb.

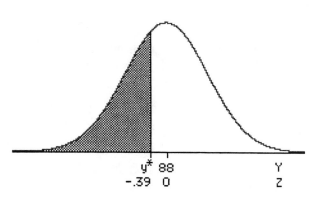

12

4.12 The distribution of readings is a normal distribution with mean μ (the true value) and standard deviation $\sigma = .008\mu$ (.8% of the true value).

(a) $\mu = 5,000,000$ which means that
$\sigma = (.008)(5,000,000) = 40,000$.
For $y = 4,900,000$,
$$z = \frac{4900000 - 5000000}{40000} = -2.5.$$
For $y = 5,100,000$,
$$z = \frac{5100000 - 5000000}{40000} = 2.5.$$
Thus, $\Pr\{4,900,000 < Y < 5,100,000\}$
$= \Pr\{-2.5 < Z < 2.5\}$
$= .9938 - .0062 = .9876$.

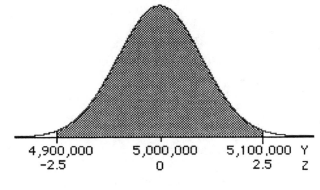

4,900,000   5,000,000   5,100,000 Y
-2.5   0   2.5   Z

(b) $\Pr\{.98\mu < Y < 1.02\mu\} = \Pr\left\{\dfrac{.98\mu - \mu}{.008\mu} < \dfrac{Y - \mu}{\sigma} < \dfrac{1.02\mu - \mu}{.008\mu}\right\}$
$= \Pr\{-2.5 < Z < 2.5\} = .9938 - .0062 = .9876$

(c) A specimen reading Y differs from the correct value by 2% or more if it does <u>not</u> satisfy $.98\mu < Y < 1.02\mu$. Using the answer from part (b), this probability is $1 - .9876 = .0124$ or 1.24%.

4.24 $\mu = 7.8$; $\sigma = 2.3$.

(a) Applying continuity correction, we wish to find the area under the normal curve to the left of $6 + .5 = 6.5$.
For $y = 6.5$,
$$z = \frac{6.5 - 7.8}{2.3} = -.57.$$
From Table 3, the area is .2843.
Thus, $\Pr\{6 \le Y\} \approx .2843$.

(b) Applying continuity correction, we wish to find the area under the normal curve between
$6 - .5 = 5.5$ and $6 + .5 = 6.5$.
For $y = 6.5$,
$$z = \frac{6.5 - 7.8}{2.3} = -.57.$$
From Table 3, the area is .2843.
For $y = 5.5$,
$$z = \frac{5.5 - 7.8}{2.3} = -1.00.$$
From Table 3, the area is .1587.
Thus,
$\Pr\{Y = 6\} \approx .2843 - .1587 = .1256$.

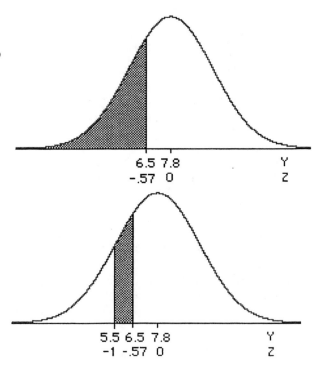

6.5 7.8   Y
-.57 0   Z

5.5 6.5 7.8   Y
-1 -.57 0   Z

(c) Applying continuity correction, we wish to find the area under the normal curve between
8 - .5 = 7.5 and 11 + .5 = 11.5.
For y = 11.5,
$$z = \frac{11.5 - 7.8}{2.3} = 1.61.$$
From Table 3, the area is .9463.
For y = 7.5,
$$z = \frac{7.5 - 7.8}{2.3} = -.13.$$
From Table 3, the area is .4483.
Thus,
$$Pr\{8 \le Y \le 11\} \approx .9463 - .4483 = .4980.$$

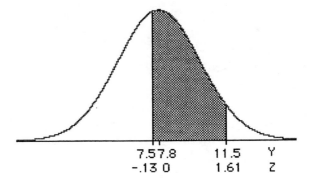

4.29 $\mu = 145$; $\sigma = 22$.

(a) For y = 100,
$$z = \frac{y - \mu}{\sigma} = \frac{100 - 145}{22} = -2.05.$$
From Table 3, the area below -2.05 is .0202.
Thus, $Pr\{Y \ge 100\} = 1 - .0202 = .9798$ or 97.98%.

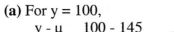

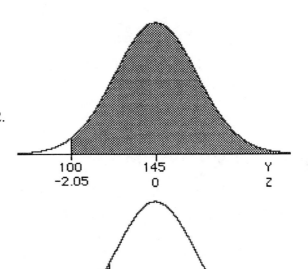

(b) For y = 120,
$$z = \frac{y - \mu}{\sigma} = \frac{120 - 145}{22} = -1.14.$$
From Table 3, the area below -1.14 is .1271.
Thus, $Pr\{Y \le 120\} = .1271$ or 12.71%.

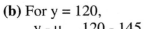

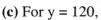

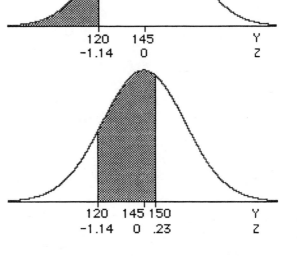

(c) For y = 120,
$$z = \frac{y - \mu}{\sigma} = \frac{120 - 145}{22} = -1.14.$$
From Table 3, the area below -1.14 is .1271.
For y = 150,
$$z = \frac{y - \mu}{\sigma} = \frac{150 - 145}{22} = .23.$$
From Table 3, the area below .23 is .5910.
Thus, $Pr\{120 \le Y \le 150\}=$
.5910 - .1271 = .4639 or 46.39%.

14

(d) For y = 100,

$$z = \frac{y - \mu}{\sigma} = \frac{100 - 145}{22} = -2.05.$$

From Table 3, the area below -2.05 is .0202.
For y = 120,

$$z = \frac{y - \mu}{\sigma} = \frac{120 - 145}{22} = -1.14.$$

From Table 3, the area below -1.14 is .1271.
Thus, Pr{100 ≤ Y ≤ 120}=
.1271 - .0202 = .1069 or 10.69%.

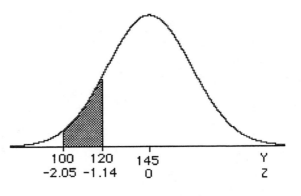

(e) For y = 150,

$$z = \frac{y - \mu}{\sigma} = \frac{150 - 145}{22} = .23.$$

From Table 3, the area below .23 is .5910.
For y = 180,

$$z = \frac{y - \mu}{\sigma} = \frac{180 - 145}{22} = 1.59.$$

From Table 3, the area below 1.59 is .9441.
Thus, Pr{150 ≤ Y ≤ 180}=
.9441 - .5910 = .3531 or 35.31%.

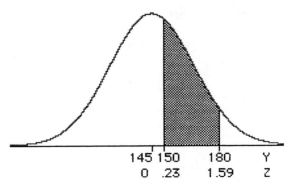

(f) For y = 180,

$$z = \frac{y - \mu}{\sigma} = \frac{180 - 145}{22} = 1.59.$$

From Table 3, the area below 1.59 is .9441.
Thus, Pr{Y ≥ 150}= 1 - .9441 = .0559 or
5.59%.

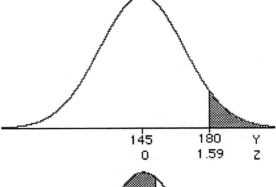

(g) For y = 150,

$$z = \frac{y - \mu}{\sigma} = \frac{150 - 145}{22} = .23.$$

From Table 3, the area below .23 is .5910.
Thus, Pr{Y ≤ 150}= .5910 or 59.10%.

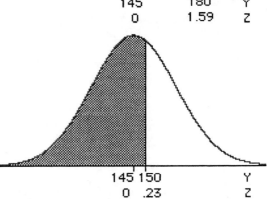

4.30 $\mu = 145$; $\sigma = 22$.

If none of the plants is more than 150 cm tall, then all of the plants are less than or equal to 150 cm tall. For y = 150

$$z = \frac{150 - 145}{22} = .23.$$

From Table 3, the area below .23 is .5910. Thus, $\Pr\{Y \le 150\} = .5910$.
We can apply the binomial formula with n = 4, p = .5910, and j = 4.
Thus, $\Pr\{\text{none more than 150 cm tall}\}$
= $\Pr\{\text{all less than or equal to 150 cm tall}\}$
= $_4C_4.5910^4 = .122.$

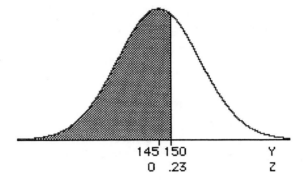

145 150   Y
0 .23   Z

4.31 $\mu = 145$; $\sigma = 22$.

The 90th percentile is the value that is larger than 90% of the distribution. In Table 3, the area closest to .9 is .8997, corresponding to z = 1.28. Thus, the 90th percentile y* satisfies the equation

$$1.28 = \frac{y^* - 145}{22}.$$

The solution of this equation is
    y* = (22)(1.28) + 145 = 173.2.
Thus, the 90th percentile of the distribution is 173.2 cm.

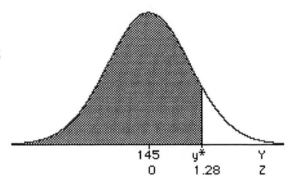

145   y*   Y
0   1.28   Z

4.33 (except μ, σ not given)

We wish to find z* such that the shaded area is .95. This means that the area in the left tail is .025. In Table 3, the area .025 corresponds to z = -1.96, which is -z*. Likewise, the area .975 corresponds to z = 1.96. Thus, z* = 1.96.

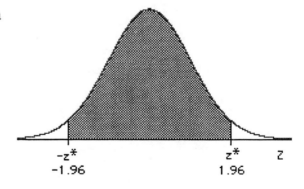

-z*   z*   Z
-1.96   1.96

4.40 $\mu = 145$; $\sigma = 22$. The distribution of readings is a normal distribution with mean μ (the true concentration) and standard deviation σ. A reading of 40 or more is considered "unusually high." Suppose that $\mu = 35$ and $\sigma = 4$.

For $y = 40$,
$$z = \frac{40 - 35}{4} = 1.25.$$

From Table 3, the area below 1.25 is .8944, which means that the area above 1.25 is $1 - .8944 = .1056$. Thus,
Pr{specimen is flagged as "unusually high}
$= .1056$.

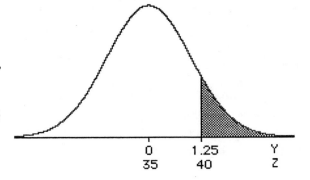

CHAPTER 5

Sampling Distributions

5.4 We are concerned with the sampling distribution of $\hat{p}$, which is governed by a binomial distribution. Letting "success" = "responder," we have $p = .2$ and $1 - p = .8$. The number of trials is $n = 15$.

   (a) The event $\hat{p} = .2$ occurs if there are 3 successes in the 15 trials (because $3/15 = .2$). Thus, to find the probability that $\hat{p} = .2$, we can use the binomial formula $_nC_j p^j (1 - p)^{n-j}$ with $j = 3$, so $n - j = 12$:
$$\Pr\{\hat{p} = .2\} = {_{15}C_3} p^3 (1 - p)^{12} = (455)(.2^3)(.8^{12}) = .2501.$$

   (b) The event $\hat{p} = 0$ occurs if there are 0 successes in the 15 trials (because $0/15 = 0$). Thus, to find the probability that $\hat{p} = 0$, we can use the binomial formula with $j = 0$, so $n - j = 15$:
$$\Pr\{\hat{p} = 0\} = {_{15}C_0} p^0 (1 - p)^{15} = (1)(1)(.8^{15}) = .0352.$$

5.5 (a) Letting "success" = "infected," we have $p = .25$ and $1 - p = .75$. The number of trials is $n = 4$. We then use the binomial formula $_nC_j p^j (1 - p)^{n-j}$ with $n = 4$ and $p = .25$. The values of $\hat{p}$ correspond to numbers of successes and failures as follows:

$\hat{p}$	Number of successes (j)	Number of failures (n - j)
$0 = 0/4$	0	4
$.25 = 1/4$	1	3
$.50 = 2/4$	2	2
$.75 = 3/4$	3	1
$1 = 4/4$	4	0

Thus, we find

(i)	$\Pr\{\hat{p} = 0\}$	$= {_4C_0} p^0 (1 - p)^4$	$= (1)(1)(.75^4)$	$= .3164$
(ii)	$\Pr\{\hat{p} = .25\}$	$= {_4C_1} p^1 (1 - p)^3$	$= (4)(.25)(.75^3)$	$= .4219$
(iii)	$\Pr\{\hat{p} = .50\}$	$= {_4C_2} p^2 (1 - p)^2$	$= (6)(.25^2)(.75^2)$	$= .2109$
(iv)	$\Pr\{\hat{p} = .75\}$	$= {_4C_3} p^3 (1 - p)^1$	$= (4)(.25^3)(.75)$	$= .0469$
(v)	$\Pr\{\hat{p} = 1\}$	$= {_4C_4} p^4 (1 - p)^0$	$= (1)(.25^4)(1)$	$= .0039$

5.9 Because $p = .40$, the event E occurs if $\hat{p}$ is within $\pm .05$ of $.40$; this happens if there are 7, 8, or 9 successes, as follows:

Number of successes (j)	$\hat{p}$
7	.35
8	.40
9	.45

18

We can calculate the probabilities of these outcomes using the binomial formula with n = 20 and p = .4:

$Pr\{\hat{p} = .35\} = {}_{20}C_7 p^7(1 - p)^{13} = (77,520)(.4^7)(.6^{13}) = .1659$

$Pr\{\hat{p} = .40\} = {}_{20}C_8 p^8(1 - p)^{12} = (125,970)(.4^8)(.6^{12}) = .1797$

$Pr\{\hat{p} = .45\} = {}_{20}C_9 p^9(1 - p)^{11} = (167,960)(.4^9)(.6^{11}) = .1597$

Finally, we calculate Pr{E} by adding these results:
  Pr{E} = .1659 + .1797 + .1597 = .5053

5.15

(a) In the population, $\mu = 176$ and $\sigma = 30$.
For y = 186,

$z = \dfrac{y - \mu}{\sigma} = \dfrac{186 - 176}{30} = .33.$

From Table 3, the area below .33 is .6293.
For y = 166,

$z = \dfrac{y - \mu}{\sigma} = \dfrac{166 - 176}{30} = -.33.$

From Table 3, the area below -.33 is .3707.
Thus, the percentage with $166 \leq y \leq 186$ is .6293 - .3707 = .2586, or 25.86%.

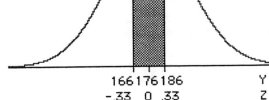

(b) We are concerned with the sampling distribution of $\bar{Y}$ for n = 9. From Theorem 5.1, the

mean of the sampling distribution of $\bar{Y}$ is
  $\mu_{\bar{Y}} = \mu = 176,$

the standard deviation is

  $\sigma_{\bar{Y}} = \dfrac{\sigma}{\sqrt{n}} = \dfrac{30}{\sqrt{9}} = 10,$

and the shape of the distribution is normal because the population distribution is normal (part 3a of Theorem 5.1).

We need to find the shaded area in the figure.

For $\bar{y} = 186,$

$z = \dfrac{\bar{y} - \mu_{\bar{Y}}}{\sigma_{\bar{Y}}} = \dfrac{186 - 176}{10} = 1.00.$

From Table 3, the area below 1.00 is .8413.
For $\bar{y} = 166,$

$z = \dfrac{\bar{y} - \mu_{\bar{Y}}}{\sigma_{\bar{Y}}} = \dfrac{166 - 176}{10} = -1.00.$

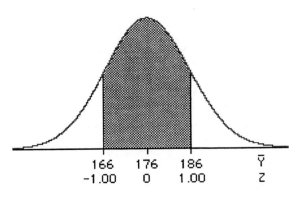

From Table 3, the area below -1.00 is .1587.

Thus, the percentage with $166 \leq \bar{y} \leq 186$ is .8413 - .1587 = .6826, or 68.26%.

(c) The probability of an event can be interpreted as the long-run relative frequency of occurrence of the event (Section 3.3). Thus, the question in part (c) is just a rephrasing of the question in part (b). It follows from part (b) that

$$\Pr\{166 \le \bar{Y} \le 186\} = .6826.$$

5.16 (a) $\mu = 3000$; $\sigma = 400$.

The event E occurs if $\bar{Y}$ is between 2900 and 3100. We are concerned with the sampling distribution of $\bar{Y}$ for n = 15. From Theorem 5.1, the mean of the sampling distribution of $\bar{Y}$ is

$$\mu_{\bar{Y}} = \mu = 3000,$$

the standard deviation is

$$\sigma_{\bar{Y}} = \frac{\sigma}{\sqrt{n}} = \frac{400}{\sqrt{15}} = 103.3,$$

and the shape of the distribution is normal because the population distribution is normal (part 3a of Theorem 5.1).

For $\bar{y} = 3100$,

$$z = \frac{\bar{y} - \mu_{\bar{Y}}}{\sigma_{\bar{Y}}} = \frac{3100 - 3000}{103.3} = .97.$$

From Table 3, the area below .97 is .8340.

For $\bar{y} = 2900$,

$$z = \frac{\bar{y} - \mu_{\bar{Y}}}{\sigma_{\bar{Y}}} = \frac{2900 - 3000}{103.3} = -.97.$$

From Table 3, the area below -.97 is .1660.

Thus, $\Pr\{2900 \le \bar{Y} \le 3100\}$
$= \Pr\{E\} = .8340 - .1660 = .6680.$

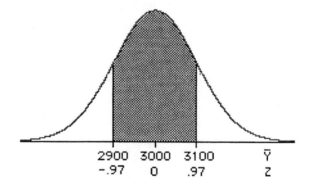

5.20 (a) In the population, 65.68% of the fish are between 51 and 60 mm long. To find the probability that four randomly chosen fish are all between 51 and 60 mm long, we let "success" be "between 51 and 60 mm long" and use the binomial distribution with n = 4 and p = .6568, as follows:

$\Pr\{$all 4 are between 51 and 60$\} = {}_4C_4 p^4 (1 - p)^0 = (1).6568^4(1) = .1861.$

(b) The mean length of four randomly chosen fish is $\bar{Y}$. Thus, we are concerned with the sampling distribution of $\bar{Y}$ for a sample of size n = 4 from a population with $\mu = 54$ and s = 4.5. From Theorem 5.1, the mean of the sampling distribution of $\bar{Y}$ is

$$\mu_{\bar{Y}} = \mu = 54,$$

the standard deviation is

$$\sigma_{\bar{Y}} = \frac{\sigma}{\sqrt{n}} = \frac{4.5}{\sqrt{4}} = 2.25,$$

and the shape of the distribution is normal because the population distribution is normal (part 3a of Theorem 5.1).

For $\bar{y} = 60$,

$$z = \frac{\bar{y} - \mu_{\bar{Y}}}{\sigma_{\bar{Y}}} = \frac{60 - 54}{2.25} = 2.67.$$

From Table 3, the area below 2.67 is .9962.

For $\bar{y} = 51$,

$$z = \frac{\bar{y} - \mu_{\bar{Y}}}{\sigma_{\bar{Y}}} = \frac{51 - 54}{2.25} = -1.33.$$

From Table 3, the area below -1.33 is .0918.

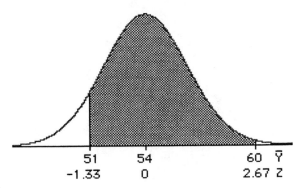

Thus, $\Pr\{51 \le \bar{Y} \le 60\}$
= .9962 - .0918 = .9044.

5.23 The distribution of repeated assays of the patient's specimen is a normal distribution with mean $\mu = 35$ (the true concentration) and standard deviation $\sigma = 4$.

(a) The result of a single assay is like a random observation Y from the population of assays. A value $Y \ge 40$ will be flagged as "unusually high." For $y = 40$,

$$z = \frac{y - \mu}{\sigma} = \frac{40 - 35}{4} = 1.25.$$

From Table 3, the area below 1.25 is .8944, so the area beyond 1.25 is
$1 - .8944 = .1056.$

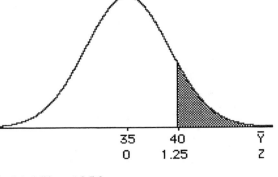

Thus, $\Pr\{$specimen will be flagged as "unusually high"$\} = .1056$.

(b) The reported value is the mean of three independent assays, which is like the mean $\bar{Y}$ of a sample of size n = 3 from the population of assays. A value $\bar{Y} \ge 40$ will be flagged as "unusually high." We are concerned with the sampling distribution of $\bar{Y}$ for a sample of size n = 3 from a population with mean $\mu = 35$ and standard deviation $\sigma = 4$. From Theorem 5.1, the mean of the sampling distribution of $\bar{Y}$ is
$$\mu_{\bar{Y}} = \mu = 35,$$

the standard deviation is

$$\sigma_{\bar{Y}} = \frac{\sigma}{\sqrt{n}} = \frac{4}{\sqrt{3}} = 2.309,$$

and the shape of the distribution is normal because the population distribution is normal (part 3a of Theorem 5.1).

For $\bar{y} = 40$,

$$z = \frac{\bar{y} - \mu_{\bar{Y}}}{\sigma_{\bar{Y}}} = \frac{40 - 35}{2.309} = 2.17.$$

From Table 3, the area below 2.17 is .9850, so the area beyond 2.17 is
$1 - .9850 = .0150.$

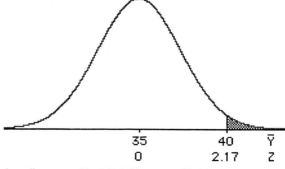

Thus, Pr{mean of three assays will be flagged as "unusually high"}= $1 - .9850 = .0150$.

5.29 For each thrust, the probability is .9 that the thrust is good and the probability is .1 that the thrust is fumbled. Letting "success" = "good thrust," and assuming that the thrusts are independent, we apply the binomial formula with $n = 4$ and $p = .9$.

(a) The area under the first peak is approximately equal to the probability that all four thrusts are good. To find this probability, we set $j = 4$; thus, the area is approximately

$$_4C_4 p^4 (1 - p)^0 = (1)(.9^4)(1) = .66.$$

(b) The area under the second peak is approximately equal to the probability that three thrusts are good and one is fumbled. To find this probability, we set $j = 3$; thus, the area is approximately

$$_4C_3 p^3 (1 - p)^1 = (4)(.9^3)(.1) = .29.$$

5.32 Letting "success" = "heads," the probability of ten heads and ten tails is determined by the binomial distribution with $n = 20$ and $p = .5$.

(a) We apply the binomial formula with $j = 10$:
Pr{10 heads, 10 tails} = $_{20}C_{10} p^{10} (1 - p)^{10} = (184,756)(.5^{10})(.5^{10}) = .1762.$

(b) According to part (a) of Theorem 5.2, the binomial distribution can be approximated by a normal distribution with
mean = $np = (20)(.5) = 10$
and
standard deviation = $\sqrt{np(1 - p)} = \sqrt{(20)(.5)(.5)} = 2.236.$
Applying continuity correction, we wish to find the area under the normal curve between
$10 - .5 = 9.5$ and $10 + .5 = 10.5$.

The desired area is shaded in the figure.

The boundary 10.5 corresponds to
$$z = \frac{10.5 - 10}{2.236} = .22.$$
From Table 3, the area below .22 is .5871.

The boundary 9.5 corresponds to
$$z = \frac{9.5 - 10}{2.236} = -.22.$$

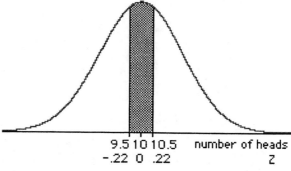

From Table 3, the area below -.22 is .4129.
Thus, the normal approximation to the binomial probability is
Pr{10 heads, 10 tails} $\approx .5871 - .4129 = .1742.$

5.34 (a) Because p = .12, the event that $\hat{p}$ will be within ±.03 of p is the event

$$.09 \leq \hat{p} \leq .15,$$

which, if n = 100, is equivalent to the event

$$9 \leq \text{number of success} \leq 15.$$

Letting "success" = "oral contraceptive user," the probability of this event is determined by the binomial distribution with

$$\text{mean} = np = (100)(.12) = 12$$

and

$$\text{standard deviation} = \sqrt{np(1 - p)} = \sqrt{(100)(.12)(.88)} = 3.250.$$

Applying continuity correction, we wish to find the area under the normal curve between 9 - .5 = 8.5 and 15 + .5 = 15.5.

The desired area is shaded in the figure.

The boundary 15.5 corresponds to

$$z = \frac{15.5 - 12}{3.250} = 1.08.$$

From Table 3, the area below 1.08 is .8599.

The boundary 8.5 corresponds to

$$z = \frac{8.5 - 12}{3.250} = -1.08.$$

From Table 3, the area below -1.08 is .1401.

Thus, the normal approximation to the binomial probability is

$$\Pr\{\hat{p} \text{ will be within } \pm.03 \text{ of } p\} \approx .8599 - .1401 = .7198.$$

(Note: An alternative method of solution is to use part (b) of Theorem 5.2 rather than part (a). Such a method is illustrated in the solutions to Exercises 5.30 and 5.41.)

5.37 (a) Because p = .3, the event E, that $\hat{p}$ will be within ±.05 of p, is equivalent to

$$.25 \leq \hat{p} \leq .35.$$

The sample size is n = 40. According to part (b) of Theorem 5.2, the sampling distribution of $\hat{p}$ can be approximated by a normal distribution with

$$\text{mean } p = .3$$

and

$$\text{standard deviation} = \sqrt{\frac{p(1 - p)}{n}} = \sqrt{\frac{(.3)(.7)}{40}} = .07246.$$

To apply continuity correction, we first calculate the half-width of a histogram bar (on the $\hat{p}$ scale) as

$$(\frac{1}{2})(\frac{1}{40}) = .1025.$$

Thus, we wish to find the area under the normal curve between

$$.25 - .0125 = .2375 \text{ and } .35 + .0125 = .3625.$$

The desired area is shaded in the figure.
The boundary .3625 corresponds to
$$z = \frac{.3625 - .25}{.07246} = .86.$$
From Table 3, the area below .86 is .8051.

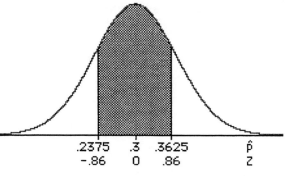

The boundary .2375 corresponds to
$$z = \frac{.2375 - .25}{.07246} = -.86.$$
From Table 3, the area below -.86 is .1949.

Thus, the normal approximation to the probability is
    $\Pr\{E\} \approx .8051 - .1949 = .6102.$

(Note: An alternative method of solution is to use part (a) of Theorem 5.2 rather than part (b). Such a method is illustrated in the solution to Exercise 5.27.)

5.42 $\mu = 88$; $\sigma = 7$.

We are concerned with the sampling distribution of $\bar{Y}$ for $n = 5$. From Theorem 5.1, the mean of the sampling distribution of $\bar{Y}$ is
    $\mu_{\bar{Y}} = \mu = 88,$

the standard deviation is
$$\sigma_{\bar{Y}} = \frac{\sigma}{\sqrt{n}} = \frac{7}{\sqrt{5}} = 3.13,$$
and the shape of the distribution is normal because the population distribution is normal (part 3a of Theorem 5.1).

For $\bar{y} = 90$,
$$z = \frac{\bar{y} - \mu_{\bar{Y}}}{\sigma_{\bar{Y}}} = \frac{90 - 88}{3.13} = .64.$$

From Table 3, the area below .64 is .7389.

Thus, $\Pr\{\bar{Y} > 90\} = 1 - .7389 = .2611.$

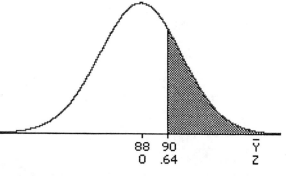

5.47 We are concerned with the sampling distribution of $\hat{p}$, which is governed by a binomial distribution. Letting "success" = "adult," we have $p = .2$ and $1 - p = .8$. The number of trials is $n = 20$.

(a) The event $\hat{p} = p$ occurs if there are 4 successes in the 20 trials (because $4/20 = .2$). Thus, to find the probability that $\hat{p} = p$, we can use the binomial formula with $j = 4$, so $n - j = 16$:
    $\Pr\{\hat{p} = p\} = {}_{20}C_4 p^4(1 - p)^{16} = (4{,}845)(.2^4)(.8^{16}) = .2182.$

(b) The event
    $p - .05 \le \hat{p} \le p + .05$
is equivalent to the event
    $.15 \le \hat{p} \le .25.$

This event occurs if there are 3, 4, or 5 successes in the 20 trials, as follows:

Number of successes (j)	$\hat{p}$
3	.25
4	.30
5	.35

We can calculate the probabilities of these outcomes using the binomial formula with $n = 20$ and $p = .2$:

$$Pr\{\hat{p} = .25\} = {}_{20}C_3 p^3 (1 - p)^{17} = (1{,}140)(.2^3)(.8^{17}) = .20536$$

$$Pr\{\hat{p} = .30\} = {}_{20}C_4 p^4 (1 - p)^{16} = (4{,}845)(.2^4)(.8^{16}) = .21820$$

$$Pr\{\hat{p} = .35\} = {}_{20}C_5 p^5 (1 - p)^{15} = (15{,}504)(.2^5)(.8^{15}) = .17456$$

Thus, $Pr\{p - .05 \le \hat{p} \le p + .05\} = Pr\{.15 \le \hat{p} \le .35\}$
$$= .20536 + .21820 + .17456 = .59812 \approx .5981.$$

5.48 We are concerned with the sampling distribution of $\hat{p}$ for $n = 20$ and $p = .2$. According to part (b) of Theorem 5.2, this sampling distribution can be approximated by a normal distribution with
mean $p = .2$
and

$$\text{standard deviation} = \sqrt{\frac{p(1 - p)}{n}} = \sqrt{\frac{(.2)(.8)}{20}} = .08944.$$

To apply continuity correction, we first calculate the half-width of a histogram bar (on the $\hat{p}$ scale) as

$$(\tfrac{1}{2})(\tfrac{1}{20}) = .025.$$

(a) We wish to find $Pr\{\hat{p} = .2\}$. Thus, we wish to find the area under the normal curve between $.2 - .025 = .175$ and $.2 + .025 = .225$.

The desired area is shaded in the figure.

The boundary .225 corresponds to
$$z = \frac{.225 - .200}{.08944} = .28.$$
From Table 3, the area below .28 is .6103.

The boundary .175 corresponds to
$$z = \frac{.175 - .200}{.08944} = -.28.$$
From Table 3, the area below -.28 is .3897.

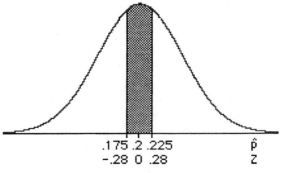

Thus, the normal approximation to the probability is
$$Pr\{\hat{p} = p\} \approx .6103 - .3897 = .2206.$$

Note that this agrees well with the exact value (.2182) found in Exercise 5.40(a).

(b) The event
$$p - .05 \le \hat{p} \le p + .05$$
is equivalent to the event
$$.15 \le \hat{p} \le .25.$$
Thus, we wish to find the area under the normal curve between
$$.15 - .025 = .125 \text{ and } .25 + .025 = .275.$$

The desired area is shaded in the figure.

The boundary .275 corresponds to
$$z = \frac{.275 - .200}{.08944} = .84.$$
From Table 3, the area below .84 is .7995.
The boundary .175 corresponds to
$$z = \frac{.125 - .200}{.08944} = -.84.$$
From Table 3, the area below -.84 is .2005.

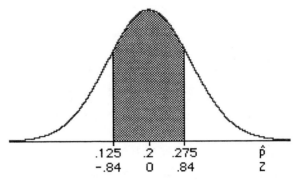

Thus, the normal approximation to the probability is
$$\Pr\{.15 \le \hat{p} \le .25\} \approx .7995 - .2005 = .5990.$$

Note that this agrees quite well with the exact value (.5981) found in Exercise 5.40(b).

5.51 $\mu = 8.3$; $\sigma = 1.7$.

If the total weight of 10 mice is 90 gm, then their mean weight is
$$\frac{90}{10} = 9.0 \text{ gm.}$$

Thus, we wish to find the percentage of litters for which $\bar{y} \ge 9.0$ gm. We are concerned with the sampling distribution of $\bar{Y}$ for $n = 10$. From Theorem 5.1, the mean of the sampling distribution of $\bar{Y}$ is
$$\mu_{\bar{Y}} = \mu = 8.3,$$
the standard deviation is
$$\sigma_{\bar{Y}} = \frac{\sigma}{\sqrt{n}} = \frac{1.7}{\sqrt{10}} = .538,$$
and the shape of the distribution is normal because the population distribution is normal (part 3a of Theorem 5.1).

We need to find the shaded area in the figure.

For $\bar{y} = 9.0$,
$$z = \frac{\bar{y} - \mu_{\bar{Y}}}{\sigma_{\bar{Y}}} = \frac{9.0 - 8.3}{.538} = 1.30.$$

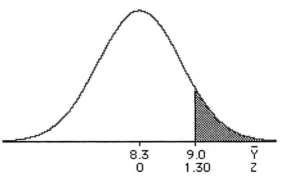

From Table 3, the area below 1.30 is .9032.

Thus, the percentage with $\bar{y} \ge 9.0$ is
$$1 - .9032 = .0968, \text{ or } 9.68\%.$$

CHAPTER 6

Confidence Intervals

6.1 (a) $\bar{y} = 1269$; $s = 145$; $n = 8$.

The standard error of the mean is

$$SE_{\bar{y}} = \frac{s}{\sqrt{n}} = \frac{145}{\sqrt{8}} = 51.3 \text{ ng/gm.}$$

(b) $\bar{y} = 1269$; $s = 145$; $n = 30$.

The standard error of the mean is

$$SE_{\bar{y}} = \frac{s}{\sqrt{n}} = \frac{145}{\sqrt{30}} = 26.5 \text{ ng/gm.}$$

6.10 (a) $\bar{y} = 31.7$ mg; $s = 8.7$ mg; $n = 5$.

The standard error of the mean is

$$SE_{\bar{y}} = \frac{s}{\sqrt{n}} = \frac{8.7}{\sqrt{5}} = 3.891 \approx 3.9 \text{ mg.}$$

(b) The degrees of freedom are n - 1 = 5 - 1 = 4. The critical value is $t_{.05} = 2.132$. The 90% confidence interval for μ is

$$\bar{y} \pm t_{.05}\frac{s}{\sqrt{n}}$$

$$31.720 \pm 2.132\left(\frac{8.7}{\sqrt{5}}\right)$$

(23.4,40.0) or $23.4 < \mu < 40.0$ mg.

(c) The degrees of freedom are n - 1 = 5 - 1 = 4. The critical value is $t_{.025} = 2.776$. The 95% confidence interval for μ is

$$\bar{y} \pm t_{.025}\frac{s}{\sqrt{n}}$$

$$31.720 \pm 2.776\left(\frac{8.7}{\sqrt{5}}\right)$$

(20.9,42.5) or $20.9 < \mu < 42.5$ mg.

6.16 (a) $\bar{y} = 13.0$; $s = 12.4$; $n = 10$.

The degrees of freedom are $n - 1 = 10 - 1 = 9$. The critical value is $t_{.05} = 2.262$. The 95% confidence interval for μ is

$$\bar{y} \pm t_{.025}\frac{s}{\sqrt{n}}$$

$$13.0 \pm 2.262\left(\frac{12.4}{\sqrt{10}}\right)$$

(4.1,21.9) or $4.1 < \mu < 21.9$ pg/ml.

(b) We are 95% confident that the average drop in HBE levels from January to May in the population of all participants in physical fitness programs like the one in the study is between 4.1 and 21.9 pg/ml.

6.20 $\bar{y} = 1.20$; $s = .14$; $n = 50$.

The degrees of freedom are $50 - 1 = 49$. From Table 4 with df = 50 (the df value closest to 49) we find that $t_{.05} = 1.676$. The 90% confidence interval for μ is

$$\bar{y} \pm t_{.05}\frac{s}{\sqrt{n}}$$

$$1.20 \pm 1.676\left(\frac{.14}{\sqrt{50}}\right)$$

(1.17,1.23) or $1.17 < \mu < 1.23$ mm.

6.24 $1 - .0025 = .9975$. In Table 3, an area of .9975 corresponds to $z = 2.81$. At distribution with df = ∞ is a normal distribution; thus, $t_{.0025} = 2.81$ when df = ∞.

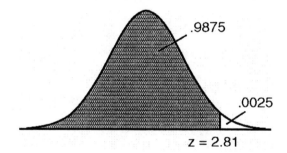

6.28 We use the inequality

$$\frac{\text{Guessed SD}}{\sqrt{n}} \leq \text{Desired SE}$$

In this case, the desired SE is 3 mg/dl and the guessed SD is 40 mg/dl. Thus, the inequality is

$$\frac{40}{\sqrt{n}} \leq 3 \text{ or } \frac{40}{3} \leq \sqrt{n} \text{ which means that } n \geq 177.8, \text{ so a sample of } n = 178 \text{ men is needed.}$$

6.37 (a) The number of mutants in the sample is $y = (100)(.20) = 20$. Thus, $\tilde{p} = (20+2)/(100+4) = .212$.

The standard error is

$$SE = \sqrt{\frac{\tilde{p}(1-\tilde{p})}{n+4}} = \sqrt{\frac{.212(1-.212)}{100+4}} = .040.$$

(b) The number of mutants in the sample is y = (400)(.20) = 80. Thus, $\tilde{p}$ = (80+2)/(400+4) = .203.

The standard error is

$$SE = \sqrt{\frac{\tilde{p}(1-\tilde{p})}{n+4}} = \sqrt{\frac{.203(1-.203)}{400+4}} = .020.$$

6.38 (a) The 95% confidence interval is

$\tilde{p} \pm 1.96 SE_{\tilde{p}}$

.212 ± (1.96)(.040)

.212 ± .078

(.134,.290) or .134 < p < .290

(b) The 95% confidence interval is

$\tilde{p} \pm 1.96 SE_{\tilde{p}}$

.203 ± (1.96)(.020)

.203 ± .039

(.164,.242) or .164 < p < .242

6.40 (a) $\tilde{p}$ = (69+2)/(339+4) = .207.

The standard error is

$$SE = \sqrt{\frac{\tilde{p}(1-\tilde{p})}{n+4}} = \sqrt{\frac{.207(1-.207)}{339+4}} = .022.$$

The 95% confidence interval is

$\tilde{p} \pm 1.96 SE_{\tilde{p}}$

.207 ± (1.96)(.022)

.207 ± .043

(.164,.250) or .164 < p < .250.

(b) We are 95% confident that the probability of adverse reaction in infants who receive their first injection of vaccine is between .164 and .250.

6.43 The required n must satisfy the inequality

$$\sqrt{\frac{(\text{Guessed } \tilde{p})(1 - \text{Guessed } \tilde{p})}{n+4}} \leq \text{Desired SE}$$

or

$$\sqrt{\frac{(.6)(.4)}{n+4}} \leq .04.$$

It follows that $\dfrac{\sqrt{(.6)(.4)}}{.04} \leq \sqrt{n+4}$

or $\dfrac{(.6)(.4)}{.04^2} \leq n+4$ or $150 \leq n+4$, so $n \geq 146$.

6.52 (a) $\bar{y} = 2.275$; $s = .238$. The standard error of the mean is

$$\mathrm{SE}_{\bar{y}} = \frac{s}{\sqrt{n}} = \frac{.238}{\sqrt{8}} = .084 \text{ mm}.$$

(b) From Table 4 with df = n - 1 = 7, we find that $t_{.025} = 2.365$.
The 95% confidence interval for μ is

$$\bar{y} \pm t_{.025}\frac{s}{\sqrt{n}}$$

$2.275 \pm (2.365)(.084)$
$(2.08, 2.47)$ or $2.08 < \mu < 2.47$ mm

(c) μ is the population mean stem diameter of plants of Tetrastichon wheat three weeks after flowering.

6.54 We use the inequality

$$\frac{\text{Guessed SD}}{\sqrt{n}} \leq \text{Desired SE}$$

In this case the desired SE is .03 mm and the guessed SD (from Exercise 6.46) is .238 mm. Thus, the inequality is

$$\frac{.238}{\sqrt{n}} \leq .03 \text{ or } \frac{.238}{.03} \leq \sqrt{n} \text{, so } 7.933^2 \leq n \text{, which means that } n \geq 62.9.$$

Thus, the experiment should include 63 plants.

6.59 (a) We must be able to view the data as a random sample of observations from a large population, the observations in the sample must be independent of each other, and the population distribution must be approximately normal. (Note, however, that because the sample size (n = 28) is not very small, some non-normality of the population distribution would be acceptable.)

(b) The shape of the histogram is an estimate of the shape of the population distribution. Thus, the histogram can be used to check the normality condition of the population

(c) If twin births were included, the independence of the observations would be questionable, because birthweights of the members of a twin pair are likely to be dependent.

6.65 $\tilde{p} = (97+2)/(123+4) = .780$.
The standard error is

$$\mathrm{SE} = \sqrt{\frac{\tilde{p}(1-\tilde{p})}{n+4}} = \sqrt{\frac{.780(1-.780)}{123+4}} = .037.$$

The 95% confidence interval is

$\tilde{p} \pm 1.96\mathrm{SE}_{\tilde{p}}$

$.780 \pm (1.96)(.037)$

$.780 \pm .073$

$(.707, .853)$ or $.707 < p < .853$.

CHAPTER 7

Comparison of Two Independent Samples

7.1 We first find the standard error of each mean.

$$SE_1 = SE_{\bar{y}_1} = \frac{s_1}{\sqrt{n_1}} = \frac{4.3}{\sqrt{6}} = 1.755.$$

$$SE_2 = SE_{\bar{y}_2} = \frac{s_2}{\sqrt{n_2}} = \frac{5.7}{\sqrt{12}} = 1.645.$$

$$SE_{(\bar{y}_1 - \bar{y}_2)} = \sqrt{SE_1^2 + SE_2^2} = \sqrt{1.755^2 + 1.645^2} = 2.41.$$

7.7 $SE_{(\bar{y}_1 - \bar{y}_2)} = \sqrt{SE_1^2 + SE_2^2} = \sqrt{.5^2 + .7^2} = .86.$

7.10 Let 1 denote males and let 2 denote females.

$\bar{y}_1 = 45.8; SE_1 = 2.8/\sqrt{489} = .127.$

$\bar{y}_2 = 40.6; SE_2 = 2.9/\sqrt{469} = .134.$

The standard error of the difference is $SE_{(\bar{y}_1 - \bar{y}_2)} = \sqrt{.127^2 + .134^2} = .185.$

The critical value $t_{.025}$ is determined from Student's t distribution with df = 950. Using df = 1000 (the nearest value given in Table 4), we find that $t(1000)_{.025} = 1.962.$

The 95% confidence interval is

$\bar{y}_1 - \bar{y}_2 \pm t_{.025}SE_{(\bar{y}_1 - \bar{y}_2)}$

$(45.8 - 40.6) \pm (1.962)(.185)$

So the confidence interval is (4.84,5.56) or $4.84 < \mu_1 - \mu_2 < 5.56.$

7.23 (a) The observed t statistic is

$$t_s = \frac{\bar{y}_1 - \bar{y}_2}{SE_{(\bar{y}_1 - \bar{y}_2)}} = \frac{735 - 854}{38} = -3.13.$$

From Table 4 with df = 4, we find the critical values $t_{.02} = 2.999$ and $t_{.01} = 3.747$. Because t_s is between $t_{.01}$ and $t_{.02}$, the P-value must be between .02 and .04. Thus, the P-value is bracketed as $.02 < P < .04.$

(b) The observed t statistic is

$$t_s = \frac{\bar{y}_1 - \bar{y}_2}{SE_{(\bar{y}_1 - \bar{y}_2)}} = \frac{5.3 - 5.0}{.24} = 1.25.$$

From Table 4 with df = 12, we find the critical values $t_{.20} = .873$ and $t_{.10} = 1.356$. Because t_s is between $t_{.10}$ and $t_{.20}$, the P-value must be between .20 and .40. Thus, the P-value is bracketed as $.20 < P < .40$.

(c) The observed t statistic is

$$t_s = \frac{\bar{y}_1 - \bar{y}_2}{SE_{(\bar{y}_1 - \bar{y}_2)}} = \frac{36 - 30}{1.3} = 4.62.$$

From Table 4 with df = 30, we find the critical value $t_{.0005} = 3.646$. Because t_s is greater than $t_{.0005}$, the P-value must be less than .001. Thus, the P-value is bracketed as $P < .001$.

7.25 (a) $.085 < .10$, which means that the P-value is less than α. Thus, we reject H_0.

(b) $.065 > .05$, which means that the P-value is greater than α. Thus, we do not reject H_0.

(c) Table 4 gives $t(19)_{.005} = 2.861$ and $t(19)_{.0005} = 3.883$, so $.001 < P < .01$. Since $P < \alpha$, we reject H_0.

(d) Table 4 gives $t(12)_{.05} = 1.782$ and $t(12)_{.04} = 1.912$, so $.08 < P < .10$. Since $P > \alpha$, we do not reject H_0.

Remark concerning tests of hypotheses The answer to a hypothesis testing exercise includes verbal statements of the hypotheses and a verbal statement of the conclusion from the test. In phrasing these statements, we have tried to capture the essence of the biological question being addressed; nevertheless the statements are necessarily oversimplified and they gloss over many issues that in reality might be quite important. For instance, the hypotheses and conclusion may refer to a causal connection between treatment and response; in reality the validity of such a causal interpretation usually depends on a number of factors related to the design of the investigation (such as unbiased allocation of animals to treatment groups) and to the specific experimental procedures (such as the accuracy of assays or measurement techniques). In short, the student should be aware that the verbal statements are intended to clarify the *statistical* concepts; their *biological* content may be open to question.

7.29 (a) The null and alternative hypotheses are

H_0: $\mu_1 = \mu_2$
H_A: $\mu_1 \neq \mu_2$

where 1 denotes heart disease and 2 denotes control. These hypotheses may be stated as

H_0: Mean serotonin concentration is the same in heart patients and in controls
H_A: Mean serotonin concentration is not the same in heart patients and in controls

The test statistic is

$$t_s = \frac{\bar{y}_1 - \bar{y}_2}{SE_{(\bar{y}_1 - \bar{y}_2)}} = \frac{3840 - 5310}{1064} = -1.38.$$

From Table 4 with df = 14, we find the critical values $t_{.10} = 1.345$ and $t_{.05} = 1.761$. Thus, the P-value is bracketed as $.10 < P < .20$.
Since the P-value is greater than α (.05), H_0 is not rejected.

(b) There is insufficient evidence $(.10 < P < .20)$ to conclude that serotonin levels are different in heart patients than in controls.

7.32 (a) The null and alternative hypotheses are

$H_0: \mu_1 = \mu_2$
$H_A: \mu_1 \neq \mu_2$

where 1 denotes flooded and 2 denotes control. These hypotheses may be stated as

H_0: Flooding has no effect on ATP
H_A: Flooding has some effect on ATP

The standard error of the difference is

$$SE_{(\bar{y}_1 - \bar{y}_2)} = \sqrt{\frac{.184^2}{4} + \frac{.241^2}{4}} = .1516.$$

The test statistic is

$$t_s = \frac{\bar{y}_1 - \bar{y}_2}{SE_{(\bar{y}_1 - \bar{y}_2)}} = \frac{1.190 - 1.785}{.1516} = -3.92.$$

From Table 4 with df $= n_1 + n_2 - 2 = 6$ (Formula (7.1) yields df $= 5.6$), we find the critical values $t_{.005} = 3.707$ and $t_{.0005} = 5.959$. Thus, the P-value is bracketed as $.001 < P < .01$. Since the P-value is less than α $(.05)$, we reject H_0.

(b) There is sufficient evidence $(.001 < P < .01)$ to conclude that flooding tends to lower ATP in birch seedlings.

7.42 If we reject H_0 (i.e., if the drug is approved) then we eliminate the possibility of a Type II error. (But by rejecting H_0 we may have made a Type I error.)

7.46 (a) The observed t statistic is

$$t_s = \frac{\bar{y}_1 - \bar{y}_2}{SE_{(\bar{y}_1 - \bar{y}_2)}} = \frac{10.8 - 10.5}{.23} = 1.30.$$

To check the directionality of the data, we note that $\bar{y}_1 > \bar{y}_2$. Thus, the data do deviate from H_0 in the direction $(\mu_1 > \mu_2)$ specified by H_A, and therefore the one-tailed P-value is the area under the t curve beyond 1.30.
From Table 4 with df $= 18$, we find the critical values $t_{.20} = .862$ and $t_{.10} = 1.330$. Because t_s is between $t_{.10}$ and $t_{.20}$, the one-tailed P-value must be between $.10$ and $.20$. Thus, the P-value is bracketed as $.10 < P < .20$.

(b) The observed t statistic is

$$t_s = \frac{\bar{y}_1 - \bar{y}_2}{SE_{(\bar{y}_1 - \bar{y}_2)}} = \frac{750 - 730}{11} = 1.82.$$

To check the directionality of the data, we note that $\bar{y}_1 > \bar{y}_2$. Thus, the data do deviate from H_0 in the direction $(\mu_1 > \mu_2)$ specified by H_A, and therefore the one-tailed P-value is the area under the t curve beyond 1.82.

From Table 4 with df = 140 (the closest value to 180), we find the critical values $t_{.04} = 1.763$ and $t_{.03} = 1.896$. Because t_s is between $t_{.03}$ and $t_{.04}$, the one-tailed P-value must be between .03 and .04. Thus, the P-value is bracketed as $.03 < P < .04$

7.48 (a) Yes. t_s is positive, as predicted by H_A. Thus, the P-value is the area under the t curve beyond 3.75. With df = 19, Table 4 gives $t_{.005} = 2.861$ and $t_{.0005} = 3.883$. Thus, $.0005 < P < .005$. Since $P < \alpha$, we reject H_0.

(b) Yes. t_s is positive, as predicted by H_A. Thus, the P-value is the area under the t curve beyond 2.6. With df = 5, Table 4 gives $t_{.025} = 2.571$ and $t_{.02} = 2.757$. Thus, $.02 < P < .025$. Since $P < \alpha$, we reject H_0.

(c) Yes. t_s is positive, as predicted by H_A. Thus, the P-value is the area under the t curve beyond 2.1. With df = 7, Table 4 gives $t_{.04} = 2.046$ and $t_{.03} = 2.241$. Thus, $.03 < P < .04$. Since $P < \alpha$, we reject H_0.

(d) No. t_s is positive, as predicted by H_A. Thus, the P-value is the area under the t curve beyond 1.8. With df = 7, Table 4 gives $t_{.10} = 1.415$ and $t_{.05} = 1.895$. Thus, $.05 < P < .10$. Since $P > \alpha$, we do not reject H_0.

7.53 The null and alternative hypotheses are

H_0: $\mu_1 = \mu_2$
H_A: $\mu_1 < \mu_2$

where 1 denotes wounded and 2 denotes control. These hypotheses may be stated as

H_0: Wounding the plant has no effect on larval growth
H_A: Wounding the plant tends to diminish larval growth

To check the directionality of the data, we note that $\bar{y}_1 < \bar{y}_2$. Thus, the data do deviate from H_0 in the direction ($\mu_1 < \mu_2$) specified by H_A. We proceed to calculate the test statistic.
The standard error of the difference is

$$SE_{(\bar{y}_1 - \bar{y}_2)} = \sqrt{\frac{9.02^2}{16} + \frac{11.14^2}{18}} = 3.46.$$

The test statistic is

$$t_s = \frac{\bar{y}_1 - \bar{y}_2}{SE_{(\bar{y}_1 - \bar{y}_2)}} = \frac{28.66 - 37.96}{3.46} = -2.69.$$

From Table 4 with df = 16 + 18 - 2 = 32 $\approx$ 30 (Formula (7.1) yields df = 31.8), we find the critical values $t_{.01} = 2.457$ and $t_{.005} = 2.750$. Thus, the P-value is bracketed as $.005 < P < .01$.
Since the P-value is less than α (.05), we reject H_0. There is sufficient evidence ($.005 < P < .01$) to conclude that wounding the plant tends to diminish larval growth.

7.54 (a) The null and alternative hypotheses are

$H_0: \mu_1 = \mu_2$

$H_A: \mu_1 > \mu_2$

where 1 denotes drug and 2 denotes placebo. These hypotheses may be stated as

H_0: The drug is not effective

H_A: The drug is effective

To check the directionality of the data, we note that $\bar{y}_1 > \bar{y}_2$. Thus, the data do deviate from H_0 in the direction $(\mu_1 > \mu_2)$ specified by H_A. We proceed to calculate the test statistic.

The standard error of the difference is

$$SE_{(\bar{y}_1 - \bar{y}_2)} = \sqrt{\frac{12.05^2}{25} + \frac{13.78^2}{25}} = 3.66.$$

The test statistic is

$$t_s = \frac{\bar{y}_1 - \bar{y}_2}{SE_{(\bar{y}_1 - \bar{y}_2)}} = \frac{31.96 - 25.32}{3.66} = 1.81.$$

From Table 4 with df = 25 + 25 -2 = 48 $\approx$ 50 (Formula (7.1) yields df = 47.2), we find the critical values $t_{.04} = 1.787$ and $t_{.03} = 1.924$. Thus, the P-value is bracketed as

$.03 < P < .04.$

Since the P-value is less than α (.05), we reject H_0. There is sufficient evidence $(.03 < P < .04)$ to conclude that the drug is effective at increasing pain relief.

(b) The only change in the calculations from part (a) would be that the one-tailed area would be doubled if the alternative were nondirectional. Thus, the p-value would be between .06 and .08 and at $\alpha = .05$ we would *not* reject H_0.

7.60 The mean difference in concentration of coumaric acid is $\mu_1 - \mu_2$, where 1 denotes dark and 2 denotes photoperiod. We construct a 95% confidence interval for $\mu_1 - \mu_2$.

$$SE_{(\bar{y}_1 - \bar{y}_2)} = \sqrt{\frac{21^2}{4} + \frac{27^2}{4}} = 17.103.$$

The critical value $t_{.025}$ is found from Student's t distribution with df = $n_1 + n_2 - 2 = 6$. (Formula (7.1) gives df = 5.7.) From Table 4, we find $t(6)_{.025} = 2.447$. The 95% confidence interval is

$\bar{y}_1 - \bar{y}_2 \pm t_{.025} SE_{(\bar{y}_1 - \bar{y}_2)}$

$(106 - 102) \pm (2.447)(17.103)$

$(-37.9, 45.9)$ or $-37.9 < \mu_1 - \mu_2 < 45.9$ nmol/gm.

The difference could be larger than 20 nmol/gm or much smaller, so the data do not indicate whether the difference is "important."

7.62 The mean difference in serum concentration of uric acid is $\mu_1 - \mu_2$, where 1 denotes men and 2 denotes women. We construct a 95% confidence interval for $\mu_1 - \mu_2$.

$$SE_{(\bar{y}_1 - \bar{y}_2)} = \sqrt{\frac{.058^2}{530} + \frac{.051^2}{420}} = .00354.$$

The critical value $t_{.025}$ is found from Student's t distribution with df $= n_1 + n_2 - 2 = 948 \approx 1000$. (Formula (7.1) gives df $= 937.8$.) From Table 4, we find $t(1000)_{.025} = 1.962$. The 95% confidence interval is

$$\bar{y}_1 - \bar{y}_2 \pm t_{.025} SE_{(\bar{y}_1 - \bar{y}_2)}$$

$$(.354 - .263) \pm (1.962)(.00354)$$

$$(.0841, .0979) \text{ or } .0841 < \mu_1 - \mu_2 < .0979 \text{ mmol/l}.$$

All values in the confidence interval are greater than .08 mmol/l. Therefore, according to the confidence interval the data indicate that the difference is "clinically important."

7.64 From the preliminary data, we obtain .3 cm as a guess of σ.
(a) If the true difference is .25 cm, then the effect size is

$$\frac{|\mu_1 - \mu_2|}{\sigma} = \frac{.25}{.3} = .83.$$

We consult Table 5 for a two-tailed test at $\alpha = .05$ and an effect size of $.83 \approx .85$; to achieve power .80, Table 5 recommends n = 23.

(b) If the true difference is .5 cm, then the effect size is

$$\frac{|\mu_1 - \mu_2|}{\sigma} = \frac{.5}{.3} = 1.67.$$

We consult Table 5 for a two-tailed test at $\alpha = .05$ and an effect size of $1.67 \approx 1.7$; to achieve power .95, Table 5 recommends n = 11.

7.67 We need to find n to achieve a power of .9. The effect size is

$$\frac{|\mu_1 - \mu_2|}{\sigma} = \frac{81 - 75}{11} = .55.$$

We consult Table 5.
(a) For a two-tailed test at $\alpha = .05$, Table 5 gives n = 71.
(b) For a two-tailed test at $\alpha = .01$, Table 5 gives n = 101.
(c) For a one-tailed test at $\alpha = .05$, Table 5 gives n = 58.

7.69 The effect size is

$$\frac{|\mu_1 - \mu_2|}{\sigma} = \frac{4}{10} = .4.$$

From Table 5 we find that, for a one-tailed test with n = 35 at significance level $\alpha = .05$, the power is .50 if the effect size is .4. Thus, the probability that Jones will reject H_0 is equal to .50.

7.77 We consult Table 6 with n = 7 and n' = 5.

    (a) $U_s = 26$. From Table 6, the smallest critical value is 27, which corresponds to a nondirectional P-value of .20. Since $U_s < 27$, it follows that $P > .20$.

    (b) $U_s = 30$. From Table 6, the critical value 30 is under the .05 heading for a nondirectional alternative. The P-value is between this heading and the next one to the right. Thus, $.02 < P .05$.

    (c) $U_s = 35$. From Table 6, the largest critical value is 34, which is under the .01 heading for a nondirectional alternative. There is no entry under the .002 heading, which means that the P-value is greater than .002. It follows that $.002 < P < .01$.

7.79 (a) The null and alternative hypotheses are

    H_0: Toluene has no effect on dopamine in rat striatum

    H_A: Toluene has some effect on dopamine in rat striatum

Let 1 denote toluene and let 2 denote control. The ordered arrays of observations are as follows:

Y_1:					1911		2314	2464		2781	2803	3420
Y_2:	1397	1803	1820	1843		1990			2539			

For the K_1 count, we note that there are four Y_2's less than the first Y_1; there are five Y_2's less than the second Y_1; there are five Y_2's less than the third Y_1; and there are six Y_2's less than the fourth, fifth, and sixth Y_1. Thus,

    $K_1 = 4 + 5 + 5 + 6 + 6 + 6 = 32$.

For the K_2 count, we note that there are no Y_1's less than the first, second, third, or fourth Y_2; there is one Y_1 less than the fifth Y_2; and there are three Y_1's less than the sixth Y_2. Thus,

    $K_2 = 0 + 0 + 0 + 0 + 1 + 3 = 4$.

To check the counts, we verify that

    $K_1 + K_2 = 32 + 4 = 36 = (6)(6) = (n_1)(n_2)$.

The Wilcoxon-Mann-Whitney test statistic is the larger of the two counts K_1 and K_2; thus $U_s = 32$.

Looking in Table 6 under n = 6 and n' = 6, we find that for a nondirectional alternative, the .05 entry is 31 and the .02 entry is 33. Thus, the P-value is bracketed as

    $.02 < P < .05$.

At significance level $\alpha = .05$, we reject H_0, since $P < .05$. We note that K_1 is larger than K_2, which indicates a tendency for the Y_1's to be larger than the Y_2's. Thus, there is sufficient evidence $(.02 < P < .05)$ to conclude that toluene increases dopamine in rat striatum.

7.86 The null and alternative hypotheses are

H_0: Mean platelet calcium is the same in people with high blood pressure as in people with normal blood pressure ($\mu_1 = \mu_2$)

H_A: Mean platelet calcium is different in people with high blood pressure than in people with normal blood pressure ($\mu_1 \neq \mu_2$)

The standard error of the difference is

$$SE_{(\bar{y}_1 - \bar{y}_2)} = \sqrt{\frac{31.7^2}{45} + \frac{16.1^2}{38}} = 5.399.$$

The test statistic is

$$t_s = \frac{\bar{y}_1 - \bar{y}_2}{SE_{(\bar{y}_1 - \bar{y}_2)}} = \frac{168.2 - 107.9}{5.399} = 11.2.$$

From Table 4 with df = 45 + 38 -2 = 81 $\approx$ 80, we find the critical value $t_{.0005} = 3.416$. The tail area is doubled for the nondirectional test. Thus, the P-value is bracketed as $P < .001$. (Formula (7.1) yields df = 67.5, but the P-value is still bracketed as $P < .001$.) Since the P-value is less than α (.01), we reject H_0. There is sufficient evidence ($P < .001$) to conclude that mean platelet calcium is higher in people with high blood pressure than in people with normal blood pressure.

7.87 The mean difference in blood pressure is $\mu_1 - \mu_2$, where 1 denotes high blood pressure and 2 denotes normal blood pressure.

The standard error of the difference is

$$SE_{(\bar{y}_1 - \bar{y}_2)} = \sqrt{\frac{31.7^2}{45} + \frac{16.1^2}{38}} = 5.399.$$

The critical value $t_{.025}$ is found from Student's t distribution with df given by Formula (7.1) as df = 67.5 $\approx$ 70. Table 4 gives $t(70)_{.025} = 1.994$. The 95% confidence interval is

$$\bar{y}_1 - \bar{y}_2 \pm t_{.025} SE_{(\bar{y}_1 - \bar{y}_2)}$$
$$(168.2 - 107.9) \pm (1.994)(5.399)$$

So the confidence interval is (49.5, 71.1) or $49.5 < \mu_1 - \mu_2 < 71.1$ nM.

Alternatively, we could use df = 45 + 38 -2 = 81 $\approx$ 80, in which case the critical value is $t(80)_{.025} = 1.990$. This gives an interval of

$$(168.2 - 107.9) \pm (1.990)(5.399)$$

So the confidence interval is (49.6, 71.0) or $49.6 < \mu_1 - \mu_2 < 71.0$ nM.

7.92 The null and alternative hypotheses are

H_0: Stress has no effect on growth

H_A: Stress tends to retard growth

The data are already arrayed in increasing order. We let Y_1 denote control and Y_2 denote stress. For the K_1 count, we note that there is one Y_2 less than the first Y_1; there are ten

Y_2's less than the second Y_1; there are twelve Y_2's less than the third, fourth, fifth, sixth, and seventh Y_1; there are twelve Y_2's less than the eighth Y_1 and one equal to it; and there are thirteen Y_2's less than the ninth through thirteenth Y_1. Thus,

$$K_1 = 1 + 10 + 12 + 12 + 12 + 12 + 12 + 12.5 + 13 + 13 + 13 + 13 + 13 = 148.5.$$

For the K_2 count, we note that there are no Y_1's less than the first Y_2; there is one Y_1 less than the second through tenth Y_2; there are two Y_1's less than the eleventh and twelfth Y_2; and there are seven Y_1's less than the thirteenth Y_2 and one equal to it. Thus,

$$K_2 = 0 + 1 + 1 + 1 + 1 + 1 + 1 + 1 + 1 + 1 + 2 + 2 + 7.5 = 20.5.$$

To check the counts, we verify that

$$K_1 + K_2 = 148.5 + 20.5 = 169 = (13)(13) = (n_1)(n_2).$$

To check the directionality of the data, we note that $K_1 > K_2$, which suggests a tendency for the Y_1's to be larger than the Y_2's, which would indicate that stress retards growth. Thus, the data do deviate from H_0 is the direction specified by H_A.

The Wilcoxon-Mann-Whitney test statistic is the larger of the two counts K_1 and K_2; thus $U_s = 148.5$.

Looking in Table 6 under n = 13 and n' = 13, we find that for a directional alternative, the largest entry is 146 under the .0005 heading. Thus, the P-value is bracketed as

   $P < .0005$.

At significance level $\alpha = .01$, we reject H_0, since $P < \alpha$. There is sufficient evidence $(P < .0005)$ to conclude that stress tends to retard growth.

7.105 False. The 95% confidence interval includes zero, which means that the P-value for a nondirectional test is greater than .05. Thus, we would not reject H_0 at the .05 significance level.

CHAPTER 8

Statistical Principles of Design

8.1 No, this does not mean that living in Arizona exacerbates breathing problems. To determine this, we would need to know whether breathing problems get better or worse for people in Arizona. In fact, people with respiratory problems often move to Arizona because the dry air is good for them. This would explain the association between living in Arizona and having breathing problems.

8.13 There is no single correct answer to this exercise, because it involves randomization.

To illustrate the procedure, we assign the animals identification numbers 1, 2, ..., 8. We then read random digits from a calculator or from Table 1. For instance, suppose the random digits are

2, 5, 9, 5, 6, 1, 3, 2, 0, 7, ...

Then we would assign the first three identification numbers -- that is, 2, 5, and 6 -- to treatment 1. (We ignore the digit 9 because no animal has this identification number. Likewise, we ignore the second 5 in the list because animal 5 has already been assigned to a treatment by the time we encounter this number.) Proceeding similarly, we would assign animals 1, 3, and 7 to treatment 2 and the remaining two animals (4 and 8) to treatment 3. This would give the following allocation:

Group 1: Animals 2, 5, 6

Group 2: Animals 1, 3, 7

Group 3: Animals 4, 8

8.17 There is no single correct answer to this exercise, because it involves randomization.

Within each litter (block), one animal will be allocated to each treatment. The random allocation is carried out separately for each litter.

To illustrate the procedure, we allocate the animals in litter 1. We assign the animals identification numbers 1, 2, 3, 4, 5. We then read random digits from a calculator or from Table 1. For instance, suppose the random digits are

2, 0, 1, 9, 4, 1, 7, 8, 2, 5, ...

Then we would assign animal 2 to treatment 1, animal 1 to treatment 2, animal 4 to treatment 3, animal 5 to treatment 4, and the remaining animal (3) to treatment 5. (We ignore the digits 0, 9, 7, and 8 because no animals have these identification numbers. Likewise, we ignore the second 1 in the list because animal 1 has already been assigned to a treatment by the time we encounter this number.)

After proceeding similarly (using new random digits) for each litter, we might obtain the following allocation:

Treatment	Piglet				
	Litter 1	Litter 2	Litter 3	Litter 4	Litter 5
1	2	5	2	4	5
2	1	4	1	1	2
3	4	2	5	2	4
4	5	3	3	3	3
5	3	1	4	5	1

8.21 Plan II is better. We want units within a block to be similar to each other; plan II achieves this. Under plan I the effect of rain would be confounded with the effect of a variety.

8.41 (a) The explanatory variable is treatment group membership (AZT or placebo).

(b) The response variable is HIV status of a baby (HIV-positive or HIV-negative).

(c) The women (or, one could say, their babies) are the experimental units, since AZT or placebo is given to the women.

CHAPTER 9

Comparison of Paired Samples

9.1 (a) The standard deviation of the four sample differences is given as .68. The standard error is

$$SE_{(\bar{y}_1 - \bar{y}_2)} = SE_{\bar{d}} = \frac{s_d}{\sqrt{n_d}} = \frac{.68}{\sqrt{4}} = .34.$$

9.3 Let 1 denote control and let 2 denote progesterone.

H_0: Progesterone has no effect on cAMP ($\mu_1 = \mu_2$)

H_A: Progesterone has some effect on cAMP ($\mu_1 \neq \mu_2$)

The standard error is

$$SE_{(\bar{y}_1 - \bar{y}_2)} = SE_{\bar{d}} = \frac{s_d}{\sqrt{n_d}} = \frac{.40}{\sqrt{4}} = .20.$$

The test statistic is

$$t_s = \frac{\bar{y}_1 - \bar{y}_2}{SE_{(\bar{y}_1 - \bar{y}_2)}} = \frac{\bar{d}}{SE_{\bar{d}}} = \frac{.68}{.20} = 3.4.$$

To bracket the P-value, we consult Table 4 with df = 4 - 1 = 3. Table 4 gives $t_{.025} = 3.182$ and $t_{.02} = 3.482$. Thus, the P-value is bracketed as

.04 < P < .05.

At significance level $\alpha = .10$, we reject H_0 if P < .10. Since .04 < P < .05, we reject H_0. There is sufficient evidence (.04 < P < .05) to conclude that progesterone decreases cAMP under these conditions.

9.4 (a) Let 1 denote treated side and 2 denote control side. The standard error is

$$SE_{(\bar{y}_1 - \bar{y}_2)} = \frac{s_d}{\sqrt{n_d}} = \frac{1.118}{\sqrt{15}} = .2887.$$

The critical value $t_{.025}$ is found from Student's t distribution with df = n_d - 1 = 15 - 1 = 14. From Table 4 we find that $t(14)_{.025} = 2.145$.

The 95% confidence interval is

$$\bar{d} \pm t_{.025}SE_{\bar{d}}$$

.117 ± (2.145)(.2887)

(-.50,.74) or $-.50 < \mu_1 - \mu_2 < .74$ °C.

9.14 (a) $B_s = 6$. Looking under $n_d = 9$ in Table 7, we see that there is no entry less than or equal to 6. Therefore, $P > .20$.

(b) $B_s = 7$. Looking under $n_d = 9$ in Table 7, we see that the only column with a critical value less than or equal to 7 is the column headed .20 (for a nondirectional alternative), and the next column is headed .10. Therefore, $.10 < P < .20$.

(c) $B_s = 8$. Looking under $n_d = 9$ in Table 7, we see that the rightmost column with a critical value less than or equal to 8 is the column headed .05 (for a nondirectional alternative), and the next column is headed .02. Therefore, $.02 < P < .05$.

(d) $B_s = 9$. Looking under $n_d = 9$ in Table 7, we see that the rightmost column with a critical value less than or equal to 9 is the column headed .01 (for a nondirectional alternative), and the next column is headed .002. Therefore, $.002 < P < .01$.

9.17 For the sign test, the hypotheses can be stated as

H_0: p=.5
H_A: p>.5

where p denotes the probability that the rat in the enriched environment will have the larger cortex. The hypotheses may be stated informally as

H_0: Weight of the cerebral cortex is not affected by environment
H_A: Environmental enrichment increases cortex weight

There were 12 pairs. Of these, there were 10 pairs in which the relative cortex weight was greater for the "enriched" rat than for his "impoverished" littermate; thus $N_+ = 10$ and $N_- = 2$. To check the directionality of the data, we note that

$N_+ > N_-$

Thus, the data so deviate from H_0 in the direction specified by H_A. The value of the test statistic is

B_s = larger of N_+ and N_-
   $= 10$.

Looking in Table 7, under $n_d = 12$ for a directional alternative, we see that the rightmost column with a critical value less than or equal to 10 is the column headed .025 and the next column is headed .01. Therefore, $.01 < P < .025$. At significance level $\alpha = .05$, we reject H_0 if $P < .05$. Since $P < .025$, we reject H_0. There is sufficient evidence ($.01 < P < .025$) to conclude that environmental enrichment increases cortex weight.

9.18 We have $n_d = 12$. The null distribution is a binomial distribution with n = 12 and p = .5. Since $B_s = 10$ and H_A is directional, we need to calculate the probability of 10, 11, or 12 plus (+) signs. We apply the binomial formula $_nC_j p^j (1 - p)^{n-j}$, as follows:

$j = 10, \quad n - j = 2$: $\quad (66)(.5^{10})(.5^2) \quad = \quad .01611$
$j = 11, \quad n - j = 1$: $\quad (12)(.5^{11})(.5^1) \quad = \quad .00293$
$j = 12, \quad n - j = 0$: $\quad (1)(.5^{12})(.5^0) \quad = \quad .00024$

The P-value is the sum of these probabilities:

$P = .01611 + .00293 + .00024 = .01928$.

9.24 (a) The null distribution is a binomial distribution with $n = 7$ and $p = .5$. Since $B_s = 7$ and H_A is nondirectional, we need to calculate the probability of 7 successes or of 0 successes. The probability of 7 successes is $.5^7 = .0078125$. Likewise, the probability of 0 successes is $.5^7 = .0078125$. Thus, $P = 2(.0078125) = .015625$.

(b) With $n_d = 7$, the smallest possible p-value is .015625; thus P cannot be less than .01.

9.45 The null and alternative hypotheses are

H_0: The average number of species is the same in pools as in riffles ($\mu_1 = \mu_2$)

H_A: The average numbers of species in pools and in riffles differ ($\mu_1 \neq \mu_2$)

The standard error is

$$SE_{(\bar{y}_1 - \bar{y}_2)} = SE_{\bar{d}} = \frac{s_d}{\sqrt{n_d}} = \frac{1.86}{\sqrt{15}} = .48.$$

The test statistic is

$$t_s = \frac{\bar{y}_1 - \bar{y}_2}{SE_{(\bar{y}_1 - \bar{y}_2)}} = \frac{\bar{d}}{SE_{\bar{d}}} = \frac{2.2}{.48} = 4.58.$$

To bracket the P-value, we consult Table 4 with df $= 15 - 1 = 14$. Table 4 gives $t_{.0005} = 4.140$. Thus, the P-value for the nondirectional test is bracketed as

$P < .001$.

At significance level $\alpha = .10$, we reject H_0 if $P < .10$. Since $P < .001$, we reject H_0. There is sufficient evidence ($P < .001$) to conclude that the average number of species in pools is greater than in riffles.

9.49 The null and alternative hypotheses are

H_0: Caffeine has no effect on RER ($\mu_1 = \mu_2$)

H_A: Caffeine has some effect on RER ($\mu_1 \neq \mu_2$)

We proceed to calculate the differences, the standard error of the mean difference, and the test statistic.

Subject	Placebo	Caffeine	Difference
1	105	96	9
2	119	99	20
3	92	89	3
4	97	95	2
5	96	88	8
6	101	95	6
7	94	88	6
8	95	93	2
9	98	88	10
Mean			7.33
SD			5.59

The standard error is

$$SE_{(\bar{y}_1 - \bar{y}_2)} = SE_{\bar{d}} = \frac{s_d}{\sqrt{n_d}} = \frac{5.59}{\sqrt{9}} = 1.86.$$

The test statistic is

$$t_s = \frac{\bar{y}_1 - \bar{y}_2}{SE_{(\bar{y}_1 - \bar{y}_2)}} = \frac{\bar{d}}{SE_{\bar{d}}} = \frac{7.33}{1.86} = 3.94.$$

To bracket the P-value, we consult Table 4 with df = 9 - 1 = 8. Table 4 gives $t_{.005} = 3.355$ and $t_{.0005} = 5.041$. Thus, the P-value for the nondirectional test is bracketed as

.001 < P < .01.

At significance level $\alpha = .05$, we reject H_0 if P < .05. Since P < .01, we reject H_0. To determine the directionality of departure from H_0, we note that

$\bar{d} > 0$; that is, $\bar{y}_1 > \bar{y}_2$.

There is sufficient evidence (.001 < P < .01) to conclude that caffeine tends to decrease RER under these conditions.

CHAPTER 10

Analysis of Categorical Data

Note: In hypothesis testing problems involving the χ^2 statistic, expected frequencies are shown in parentheses.

10.1 The hypotheses are

H_0: The model is correct (the population ratio is 12:3:1)

H_A: The model is incorrect (the population ratio is not 12:3:1)

More formally, we can state these as

H_0: Pr{white} = .75, Pr{yellow} = .1875, Pr{green} = .0625

H_A: At least one of the probabilities specified by H_0 is incorrect

We calculate the expected frequencies from H_0 as follows:

White:	E =	(.75)(205)	=	153.75
Yellow:	E =	(.1875)(205)	=	38.4375
Green:	E =	(.0625)(205)	=	12.8125

The observed and expected frequencies (in parentheses) are:

White	Yellow	Green	Total
155 (153.75)	40 (38.4375)	10 (12.8125)	205

The χ^2 test statistic is

$$\chi^2{}_s = \frac{(155 - 153.75)^2}{153.75} + \frac{(40 - 38.4375)^2}{38.4375} + \frac{(10 - 12.8125)^2}{12.8125} = .69.$$

There are 3 categories, so we consult Table 8 with df = 3 - 1 = 2. From Table 8, we find $\chi^2(2)_{.20} = 3.22$. Because $\chi^2{}_s < \chi^2{}_{.20}$, the P-value is bracketed as

P > .20.

At significance level .10, we would reject H_0 if P < .10. Since P > .20, we do not reject H_0. There is little or no evidence (P > .20) that the model is not correct; the data are consistent with the model.

10.2 H_0 and H_A are the same as in Exercise 10.1. Because the sample is 10 times as large, the value of $\chi^2{}_s$ is 10 times as large as in Exercise 10.1. Thus,

$$\chi^2{}_s = (10)(.69) = 6.9.$$

From Table 8, with df = 3 - 1 = 2, we find $\chi^2(2)_{.05} = 5.99$ and $\chi^2(2)_{.02} = 7.82$; thus, the P-value is bracketed as

.02 < P < .05.

At significance level .10, we reject H_0 if P < .10. Since .02 < P < .05, we reject H_0. There is sufficient evidence (.02 < P < .05) to conclude that the model is incorrect; the data are not consistent with the model. (Note that, because H_0 is a compound hypothesis, the conclusion for the χ^2 test is nondirectional.)

10.8 The hypotheses may be stated informally as

H_0: The drug does not cause tumors

H_A: The drug causes tumors

Consider only the 20 triplets in which at least one tumor occurred. Let T denote the event that a tumor occurs in the treated rat before a tumor occurs in a control rat. If the drug does not cause tumors, then each rat is equally likely to be the first to develop a tumor, so that $Pr\{T\}$ would be 1/3. On the other hand, if the drug does cause tumors, then the treated rat is at higher risk, so that $Pr\{T\}$ would be greater than 1/3. Thus, the hypotheses can be stated formally as

$$H_0: Pr\{T\} = \frac{1}{3}$$

$$H_A: Pr\{T\} > \frac{1}{3}$$

Because H_A is directional, we begin by checking the directionality of the data. The estimated probability of T is

$$\hat{Pr}\{T\} = \frac{12}{20} = .6.$$

We note that

$$\hat{Pr}\{T\} > \frac{1}{3}.$$

Thus, the data do deviate from H_0 in the direction specified by H_A. We proceed to the calculation of the test statistic. The following are the observed and expected frequencies (in parentheses):

Tumor first in treated rat	Tumor first in control rat
12 (6.67)	8 (13.33)

The expected frequencies are calculated as $\frac{1}{3}(20)$ and $\frac{2}{3}(20)$. The χ^2 statistic is

$$\chi^2_s = \frac{(12 - 6.67)^2}{6.67} + \frac{(8 - 13.33)^2}{13.33} = 6.4.$$

There are 2 categories, so we consult Table 8 with df = 2 - 1 = 1. From Table 8, we find $\chi^2(1)_{.02} = 5.41$ and $\chi^2(1)_{.01} = 6.63$, so $\chi^2(1)_{.01} < \chi^2_s < \chi^2(1)_{.02}$. Because H_A is directional, the column headings (.02 and .01) must be cut in half to bracket the P-value; thus P-value is bracketed as

$.005 < P < .01.$

At significance level .01, we reject H_0 if $P < .01$. Since $.005 < P < .01$, we reject H_0. There is sufficient evidence $(.005 < P < .01)$ to conclude that the drug does cause tumors.

10.15 (a) To have $\chi^2_s = 0$, the columns of the table (and the rows of the table) must be proportional to each other, as in the following table:

	Treatment 1	2	
Success	5	20	
Failure	10	40	
Total	15	60	

(b) The estimated probabilities of success are $\hat{p}_1 = 5/15 = 1/3$ and $\hat{p}_2 = 20/60 = 1/3$. Yes, these proportions are equal.

10.18 The hypotheses are

H_0: Mites do not induce resistance to wilt

H_A: Mites do induce resistance to wilt

Letting p denote the probability of wilt and letting 1 denote mites and 2 denote no mites, the hypotheses may be stated as

H_0: $p_1 = p_2$

H_A: $p_1 < p_2$

Because H_A is directional, we begin by checking the directionality of the data. The estimated probabilities of wilt disease are

$$\hat{p}_1 = \frac{11}{26} \approx .42$$

$$\hat{p}_2 = \frac{17}{21} \approx .81$$

We note that

$$\hat{p}_1 < \hat{p}_2.$$

Thus, the data do deviate from H_0 in the direction specified by H_A. We proceed to the calculation of the test statistic. The expected frequency for any given cell is found from the formula

$$E = \frac{(\text{Row total}) \times (\text{Column total})}{\text{Grand total}}$$

The following table shows the observed and expected frequencies (in parentheses):

	Mites	No mites	Total
Wilt disease	11 (15.49)	17 (12.51)	28
No wilt disease	15 (10.51)	4 (8.49)	19
Total	26	21	47

The χ^2 test statistic is

$$\chi^2_s = \frac{(11 - 15.49)^2}{15.49} + \frac{(17 - 12.51)^2}{12.51} + \frac{(15 - 10.51)^2}{10.51} + \frac{(4 - 8.49)^2}{8.49} = 7.21.$$

48

We consult Table 8 with df = 1. From Table 8, we find $\chi^2(1)_{.01} = 6.63$ and $\chi^2(1)_{.001} = 10.83$, so $\chi^2(1)_{.001} < \chi^2_s < \chi^2(1)_{.01}$. Because H_A is directional, the column headings (.01 and .001) must be cut in half to bracket the P-value; thus P-value is bracketed as

.0005 < P < .005.

At significance level .01, we reject H_0 if P < .01. Since .0005 < P < .005, we reject H_0. There is sufficient evidence (.0005 < P < .005) to conclude that mites do induce resistance to wilt.

10.23 Let p denote the probability of response, let 1 denote simultaneous, and let 2 denote sequential administration. The hypotheses are

H_0: The two timings are equally effective ($p_1 = p_2$)
H_A: The two timings are not equally effective ($p_1 \neq p_2$)

The expected frequency for any given cell is found from the formula

$$E = \frac{(\text{Row total}) \times (\text{Column total})}{\text{Grand total}}$$

The following table shows the observed and expected frequencies (in parentheses):

	Simultaneous	Sequential	Total
Response	11 (8.30)	3 (5.70)	14
No response	5 (7.70)	8 (5.30)	13
Total	16	11	27

The χ^2 test statistic is

$$\chi^2_s = \frac{(11 - 8.30)^2}{8.30} + \frac{(3 - 5.70)^2}{5.70} + \frac{(5 - 7.70)^2}{7.70} + \frac{(8 - 5.30)^2}{5.30} = 4.48.$$

We consult Table 8 with df = 1. From Table 8, we find $\chi^2(1)_{.05} = 3.84$ and $\chi^2(1)_{.02} = 5.41$, so $\chi^2(1)_{.02} < \chi^2_s < \chi^2(1)_{.05}$. Thus P-value is bracketed as

.02 < P < .05.

At significance level .05, we reject H_0 if P < .05. Since .02 < P < .05, we reject H_0. To determine directionality, we calculate

$$\hat{p}_1 = \frac{11}{16} \approx .69$$

$$\hat{p}_2 = \frac{3}{11} \approx .27$$

and we note that

$$\hat{p}_1 > \hat{p}_2.$$

There is sufficient evidence (.02 < P < .05) to conclude that the simultaneous timing is superior to the sequential timing.

10.29 (a) $\hat{\text{Pr}}\{D|P\} = \dfrac{29}{109} = .266$

$\hat{\text{Pr}}\{D|N\} = \dfrac{10}{104} = .096$

$\hat{\text{Pr}}\{P|D\} = \dfrac{29}{39} = .744$

$\hat{\text{Pr}}\{P|A\} = \dfrac{80}{174} = .460$

(b) The hypotheses are

H_0: There is no association between antibody and survival $(\text{Pr}\{D|P\} = \text{Pr}\{D|N\})$

H_A: There is some association between antibody and survival $(\text{Pr}\{D|P\} \neq \text{Pr}\{D|N\})$.

The test statistic is $\chi^2_s = 10.27$. With df = 1, Table 8 gives $\chi^2_{.01} = 6.63$ and $\chi^2_{.001} = 10.83$, so the P-value is bracketed as

$.001 < P < .01$.

Because $P < \alpha$, we reject H_0. There is sufficient evidence $(.001 < P < .01)$ to conclude that men with antibody are less likely to survive 6 months than men without antibody $(\text{Pr}\{D|P\} > \text{Pr}\{D|N\})$.

10.30 The data in Exercise 10.24 suggest that men with antibody are less likely to survive six months than men without antibody. If survival is predicted solely on the basis of the antibody test, then the prediction rule would be:

For men with antibody, predict death

For men without antibody, predict survival

If this prediction rule were applied to the 213 men in Exercise 10.24, the results would be as follows:

For the men with antibody, the prediction would be death; 29 of these would be correct predictions.

For the men without antibody, the prediction would be survival; 94 of these would be correct predictions.

Thus, the overall probability of a correct prediction is estimated to be

$\hat{\text{Pr}}\{\text{correct prediction}\} = \dfrac{29 + 94}{213} = .577$.

Based on this estimate, the antibody test does not appear to be a good predictor of survival.

(Note: One could argue that the prediction rule should be "Predict survival for everyone, since $\hat{\text{Pr}}\{D|P\} = .266$ and $\hat{\text{Pr}}\{D|N\} = .096$ are both less than .5." However, this rule ignores the outcome of the antibody test, so it is not in the spirit of predicting survival "solely on the basis of the antibody test.")

10.31 The following table shows the data arranged as a contingency table.

		Preferred	Hand	
		Right	Left	Total
Preferred	Right	2012	121	2133
Foot	Left	142	116	258
	Total	2154	237	2391

Let RH, LH, RF, and LF denote right-handed, left-handed, right-footed, and left-footed.

(a) The estimated conditional probability that a woman is right-footed, given that she is right-handed, is $\hat{\text{Pr}} \{RF|RH\} = \dfrac{2012}{2154} = .934$.

(b) The estimated conditional probability that a woman is right-footed, given that she is left-handed, is $\hat{\text{Pr}} \{RF|LH\} = \dfrac{121}{237} = .511$.

(c) We test independence of hand preference and foot preference using a χ^2 test. We calculated expected frequencies from the formula

$$E = \frac{(\text{Row total}) \times (\text{Column total})}{\text{Grand total}}$$

The following table shows the expected frequencies:

		Preferred	Hand	
		Right	Left	Total
Preferred	Right	1921.57	211.43	2133
Foot	Left	232.43	25.57	258
	Total	2154	237	2391

The χ^2 test statistic is

$$\chi^2_s = \frac{(2012 - 1921.57)^2}{1921.57} + \frac{(121 - 211.43)^2}{211.43} + \frac{(142 - 232.43)^2}{232.43} + \frac{(116 - 25.57)^2}{25.57} = 398.$$

(d) The null hypothesis can be expressed as

$$H_0: \text{Pr}\{RF|RH\} = .5 = \text{Pr}\{LF|RH\}$$

This requires a goodness-of-fit test. The expected frequencies are calculated from H_0 as

Right-footed: $E = (.5)(2154) = 1077$
Left-footed: $E = (.5)(2154) = 1077$

The observed frequencies are
Right-footed: 2012
Left-footed: 142

The χ^2 test statistic is

$$\chi^2_s = \frac{(2012 - 1077)^2}{1077} + \frac{(142 - 1077)^2}{1077} = 1{,}623.$$

10.40 Tables that more strongly support H_A are those with fewer than 2 deaths on treatment B. There are two such tables:

5	1
9	15

6	0
8	16

10.49 (a) Letting UP denote ulcer patient and C denote control, the hypotheses are

H_0: The blood type distributions are the same for ulcer patients and controls ($Pr\{O|UP\}$ = $Pr\{O|C\}$, $Pr\{A|UP\} = Pr\{A|C\}$, $Pr\{B|UP\} = Pr\{B|C\}$, $Pr\{AB|UP\} = Pr\{AB|C\}$)
H_A: The blood type distributions are not the same

The test statistic is $\chi^2_s = 49.0$. The degrees of freedom are df $= (4 - 1)(2 - 1) = 3$. From Table 8 we find that $\chi^2(3)_{.0001} = 21.11$. Because $\chi^2_s < \chi^2_{.0001}$, the P-value is bracketed as

$P < .0001$.

At significance level $\alpha = .01$, we will reject H_0 if $P < .01$. Since $P < .0001$, we reject H_0 and conclude that the blood type distribution of ulcer patients is different from that of controls.

10.59 $\hat{p}_1 = 32/105 = .30476$, $\hat{p}_2 = 20/107 = .18692$

$$SE_{(\hat{p}_1 - \hat{p}_2)} = \sqrt{\frac{(.30476)(.69524)}{105} + \frac{(.18692)(.81308)}{107}} = .05864$$

$(.30476 - .18692) \pm (1.96)(.05864)$

$(.003, .233)$ or $.003 < p_1 - p_2 < .233$. No; the confidence interval suggests that bed rest may actually be harmful.

10.61 (a) $\hat{p}_1 = 911/1655 = .55045$, $\hat{p}_2 = 4578/10000 = .4578$

$$SE_{(\hat{p}_1 - \hat{p}_2)} = \sqrt{\frac{(.55045)(.44955)}{1655} + \frac{(.4578)(.5422)}{10000}} = .0132$$

$(.55045 - .4578) \pm (1.96)(.0132)$

$(.067, .119)$ or $.067 < p_1 - p_2 < .119$

(b) We are 95% confident that the proportion of persons with type O blood among ulcer patients is higher than the proportion of persons with type O blood among healthy individuals by between .067 and .119. That is, we are 95% confident that p_1 exceeds p_2 by between .067 and .119.

10.63 The data are

		Case	
		No	Yes
Control	No	107	30
	Yes	13	5

Note: There is an error in the first printing of the 2nd edition, whereby the headings of "Yes" and "No" were reversed.

The hypotheses are

H_0: There is no association between oral contraceptive use and stroke (p = .5)
H_A: There is an association between oral contraceptive use and stroke (p ≠ .5)

where p denotes the probability that a discordant pair will be Yes(case)/No(control).

The test statistic for McNemar's test is $\chi^2_s = \dfrac{(13 - 30)^2}{13 + 30} = 6.72$.

Looking in Table 8, with df = 1, we see that $\chi^2_{.01} = 6.63$ and $\chi^2_{.001} = 10.83$. Thus, .001 < P < .01, so we reject H_0. There is sufficient evidence (.001 < P < .01) to conclude that stroke victims are more likely to be oral contraceptive users (p > .5).

10.66 (a) (i) $\hat{p}_1 = 25/517 = .04836$ and $\hat{p}_2 = 23/637 = .0361$. The relative risk is $\hat{p}_1/\hat{p}_2 = .04836/.03611 = 1.339$.

(ii) The odds ratio is $\dfrac{(25)(614)}{(23)(492)} = 1.356$.

(b) (i) $\hat{p}_1 = 12/105 = .11429$ and $\hat{p}_2 = 8/92 = .08696$. The relative risk is $\hat{p}_1/\hat{p}_2 = .11429/.08696 = 1.314$.

(ii) The odds ratio is $\dfrac{(12)(84)}{(8)(93)} = 1.355$.

10.72 (a) The sample odds ratio is $\dfrac{(309)(1341)}{(266)(1255)} = 1.241$.

(b) $\log(\hat{\theta}) = .2159$; $SE_{\log(\hat{\theta})} = \sqrt{\dfrac{1}{309} + \dfrac{1}{1255} + \dfrac{1}{266} + \dfrac{1}{1341}} = .092$. The 95% confidence interval for $\log(\theta)$ is $.2159 \pm (1.96)(.092)$, which is $(.036, .396)$. $e^{.036} = 1.04$ and $e^{.396} = 1.49$. The 95% confidence interval for θ is $(1.04, 1.49)$.

(c) We are 95% confident that taking heparin increases the odds of a negative response by a factor of between 1.04 and 1.49 when compared to taking enoxaparin. Since a negative outcome is fairly rare, we can say that we are 95% confident that the probability of a negative outcome is between 1.04 and 1.49 times higher for patients given heparin than for patients given enoxaparin.

10.76 Let p denote the probability of female and let 1 and 2 denote warm and cold environments.

(a) H_0: The sex ratio is 1:1 in the warm environment ($p_1 = .5$)
H_A: Sex ratio is not 1:1 in the warm environment ($p_1 \neq .5$),

The expected frequencies are calculated from H_0 as follows:

Female: E = (.5)(141) = 70.5
Male: E = (.5)(141) = 70.5

The observed and expected frequencies (in parentheses) are

Female	Male	Total
73 (70.5)	68 (70.5)	141

The χ^2 statistic is

$$\chi^2_s = \frac{(73 - 70.5)^2}{70.5} + \frac{(68 - 70.5)^2}{70.5} = .18.$$

There are two categories (female and male), so we consult Table 8 with df = 2 - 1 = 1. From Table 8, we find $\chi^2(1)_{.20} = 1.64$. Because $\chi^2_s < \chi^2_{.20}$, the P-value is bracketed as

P > .20.

At significance level $\alpha = .05$, we reject H_0 if P < .05. Since P > .20, we do not reject H_0. There is insufficient evidence (P > .20) to conclude that the sex ratio is not 1:1 in the warm environment.

(c) The hypotheses are

H_0: Sex ratio is the same in the two environments ($p_1 = p_2$)
H_A: Sex ratio is not the same in the two environments ($p_1 \neq p_2$)

We calculate the expected frequencies under H_0 from the formula

$$E = \frac{(Row\ total) \times (Column\ total)}{Grand\ total}$$

The following table shows the observed and expected frequencies (in parentheses):

	Environment		
	Warm	Cold	Total
Male	68 (59.13)	62 (70.87)	130
Female	73 (81.87)	107 (98.13)	180
Total	141	169	310

The χ^2 test statistic is

$$\chi^2_s = \frac{(68 - 59.13)^2}{59.13} + \frac{(62 - 70.87)^2}{70.87} + \frac{(73 - 81.87)^2}{81.87} + \frac{(107 - 98.13)^2}{98.13} = 4.20.$$

The degrees of freedom are df = (2 - 1)(2 - 1) = 1. Table 8 shows that $\chi^2(1)_{.05} = 3.84$ and $\chi^2(1)_{.02} = 5.41$. Thus, .02 < P < .05, so H_0 is rejected. To determine directionality, we calculate

$$\hat{p}_1 = \frac{73}{141} = .52$$

$$\hat{p}_1 = \frac{107}{169} = .63$$

and we note that $\hat{p}_1 < \hat{p}_2$.

There is sufficient evidence (.02 < P < .05) to conclude that the probability of a female is higher in the cold than the warm environment.

10.80 (a) The hypotheses are

H_0: Directional choice under cloudy skies is random ($\Pr\{\text{toward}\} = .25$, $\Pr\{\text{away}\} = .25$, $\Pr\{\text{right}\} = .25$, $\Pr\{\text{left}\} = .25$)

H_A: Directional choice under cloudy skies is not random

The expected frequencies, under H_0, are

Toward: $E = (.25)(50) = 12.5$
Away: $E = (.25)(50) = 12.5$
Right: $E = (.25)(50) = 12.5$
Left: $E = (.25)(50) = 12.5$

The observed and expected frequencies (in parentheses) are

Toward	Away	Right	Left	Total
18 (12.5)	12 (12.5)	13 (12.5)	7 (12.5)	50

The χ^2 statistic is

$$\chi^2_s = \frac{(18 - 12.5)^2}{12.5} + \frac{(12 - 12.5)^2}{12.5} + \frac{(13 - 12.5)^2}{12.5} + \frac{(7 - 12.5)^2}{12.5} = 4.88.$$

There are four categories, so we consult Table 8 with df = 4 - 1 = 3. From Table 8, we find $\chi^2(3)_{.20} = 4.64$ and $\chi^2(3)_{.10} = 6.25$. Because $\chi^2_{.20} < \chi^2_s < \chi^2_{.10}$, the P-value is bracketed as $.10 < P < .20$.

At significance level $\alpha = .05$, we reject H_0 if $P < .05$. Since $.10 < P < .20$, we do not reject H_0. There is insufficient evidence ($.10 < P < .20$) to conclude that the directional choice is not random.

10.89 The null and alternative hypotheses are

H_0: Site of capture and site of recapture are independent ($\Pr\{\text{RI}|\text{CI}\} = \Pr\{\text{RI}|\text{CII}\}$)

H_A: Flies preferentially return to their site of capture ($\Pr\{\text{RI}|\text{CI}\} > \Pr\{\text{RI}|\text{CII}\}$)

where C and R denote capture and recapture and I and II denote the sites.

Because H_A is directional, we begin by checking the directionality of the data. We calculate

$$\hat{\Pr}\{\text{RI}|\text{CI}\} = \frac{78}{134} = .58$$

$$\hat{\Pr}\{\text{RI}|\text{CII}\} = \frac{33}{91} = .36$$

and we note that $\hat{\Pr}\{\text{RI}|\text{CI}\} > \hat{\Pr}\{\text{RI}|\text{CII}\}$.

Thus, the data deviate from H_0 in the direction specified by H_A.

The test statistic is $\chi^2_s = 10.44$. From Table 8 with df = 1, we find $\chi^2(1)_{.01} = 6.63$ and $\chi^2(1)_{.001} = 10.83$. We cut the column headings in half for the directional test. Thus, $.0005 < P < .005$ and we reject H_0. At the .01 level, there is sufficient evidence ($.0005 < P < .005$) to conclude that flies preferentially return to their site of capture.

10.94 The null and alternative hypotheses are

H_0: The probability of an egg being on a particular type of bean is .25 for all four types of beans

H_A: H_0 is false (at least one of the probabilities is not .25)

Under H_0, the expected number of eggs for each type of bean is (.25)(Total), which is (.25)(711) = 177.75. The observed and expected frequencies are

Pinto	Cowpea	Navy	Northern
167 (177.75)	176 (177.75)	174 (177.75)	194 (177.75)

The test statistic is

$$\chi^2_s = \frac{(167 - 177.75)^2}{177.75} + \frac{(176 - 177.75)^2}{177.75} + \frac{(174 - 177.75)^2}{177.75} + \frac{(194 - 177.75)^2}{177.75} = 2.23.$$

There are 4 categories, so df = 4 - 1 = 3. Table 8 gives $\chi^2(3)_{.20} = 4.64$, so P > .20 and we do not reject H_0. There is insufficient evidence (P > .20) to conclude that cowpea weevils prefer one type of bean over the others.

CHAPTER 11

Comparing the Means of k Independent Samples

11.1 We have $n^* = 4 + 3 + 4 = 11$;

$$\sum_{i=1}^{k} \sum_{j=1}^{n_i} y_{ij} = 48 + 39 + 42 + 43 + 40 + 48 + 44 + 39 + 30 + 32 + 35 = 440;$$

$$\bar{\bar{y}} = \frac{440}{11} = 40.$$

(a) $SS(\text{between}) = (4)(48 - 43)^2 + (3)(44 - 40)^2 + (4)(34 - 40)^2 = 228$;

$SS(\text{within}) = (48 - 43)^2 + (39 - 43)^2 + (42 - 43)^2 + (43 - 43)^2$
$\qquad + (40 - 44)^2 + (48 - 44)^2 + (44 - 44)^2$
$\qquad + (39 - 34)^2 + (30 - 34)^2 + (32 - 34)^2 + (35 - 34)^2 = 120.$

(b) $SS(\text{total}) = (48 - 40)^2 + (39 - 40)^2 + (42 - 40)^2 + (43 - 40)^2$
$\qquad + (40 - 40)^2 + (48 - 40)^2 + (44 - 40)^2$
$\qquad + (39 - 40)^2 + (30 - 40)^2 + (32 - 40)^2 + (35 - 40)^2 = 348.$

Verification: $SS(\text{between}) + SS(\text{within}) = SS(\text{total})$;
$\qquad\qquad\qquad\quad 228 + 120 = 348.$

(c) $df(\text{between}) = k - 1 = 3 - 1 = 2$; $MS(\text{between}) = \dfrac{SS(\text{between})}{df(\text{between})} = \dfrac{228}{2} = 114$;

$df(\text{within}) = n^* - k = 11 - 3 = 8$; $MS(\text{within}) = \dfrac{MS(\text{within})}{df(\text{within})} = \dfrac{120}{8} = 15$;

$s_{\text{pooled}} = \sqrt{MS(\text{within})} = \sqrt{15} = 3.87.$

11.4 (a) We find $SS(\text{between})$ by subtraction: $SS(\text{between}) = 472 - 337 = 135.$

We find $df(\text{total})$ by adding $df(\text{between})$ and $df(\text{within})$: $df(\text{total}) = 3 + 12 = 15.$

We find $MS(\text{within})$ by division: $MS(\text{within}) = SS(\text{within})/df(\text{within}) = 337/12 = 28.08.$

The completed table is

Source	df	SS	MS
Between groups	3	135	45
Within groups	12	337	28.08
Total	15	472	

(b) We have $df(\text{between}) = 3 = k - 1$, so $k = 4.$

(c) We have $df(\text{total}) = 15 = n^* - 1$, so $n^* = 16.$

11.9 (a) The hypotheses are

H_0: The stress conditions all produce the same mean lymphocyte concentration
($\mu_1 = \mu_2 = \mu_3 = \mu_4$)

H_A: Some of the stress conditions produce different mean lymphocyte concentrations (the μ's are not all equal)

The number of groups is $k = 4$ and the total number of observations is $n^* = 48$. Thus, df(between) = $k - 1 = 3$ and df(within) = $n^* - k = 44$.

Source	df	SS	MS
Between groups	3	89.036	29.68
Within groups	44	340.24	7.733
Total	47	429.28	

The test statistic is $F_s = \dfrac{MS(between)}{MS(within)} = \dfrac{29.68}{7.733} = 3.84$. With df = 3 and 40 (the closest value to 44), Table 9 gives $F_{.02} = 3.67$ and $F_{.01} = 4.31$. Thus, we have $.01 < P < .02$. Since $P < \alpha$, we reject H_0. There is sufficient evidence ($.01 < P < .02$) to conclude that some of the stress conditions produce different mean lymphocyte concentrations.

(b) $s_{pooled} = \sqrt{MS(within)} = \sqrt{7.733} = 2.78$ cells/ml x 10^{-6}.

11.10 (a) The null hypothesis is

H_0: Mean HBE is the same in all three populations

(d) $s_{pooled} = \sqrt{MS(within)} = \sqrt{208.7} = 14.4$ pg/ml.

11.24 (a) For women with no children, the age-adjusted blood pressure is the following linear combination:
$L = (.17)(113) + (.29)(118) + (.31)(125) + (.23)(134) = 123$ mm Hg.

(b) For women with five or more children, the age-adjusted blood pressure is the following linear combination:
$L = (.17)(114) + (.29)(116) + (.31)(124) + (.23)(138) = 123.2$ mm Hg.

(d) For the linear combination in part (a), the multipliers are
.17, .29, .31, .23.

The standard error is

$$SE_L = 18\sqrt{\frac{.17^2}{230} + \frac{.29^2}{110} + \frac{.31^2}{105} + \frac{.23^2}{123}} = .851 \text{ mm Hg.}$$

<u>Note</u>: For exercises that require definition of contrasts, the answer given is not the only correct answer. For example, reversing the sign of all contrasts is also correct.

11.28 (a) The population mean difference between T and H at the high (Hi) dose is

$$\mu_{T,Hi} - \mu_{H,Hi}$$

and at the low dose is

$$\mu_{T,Lo} - \mu_{H,Lo}.$$

The average of these is

$$\frac{1}{2}(\mu_{T,Hi} - \mu_{H,Hi}) + \frac{1}{2}(\mu_{T,Lo} - \mu_{H,Lo}) \, .$$

The corresponding linear combination is

$$L = \frac{1}{2}(17.1 - 17.5) + \frac{1}{2}(13.9 - 15.8) = -1.15.$$

The m's are $\frac{1}{2}, \frac{-1}{2}, \frac{1}{2}$, and $\frac{-1}{2}$. Thus, the standard error of L is

$$SE_L = \sqrt{11.83^2 \left(\frac{(1/2)^2}{53} + \frac{(-1/2)^2}{57} + \frac{(-1/2)^2}{55} + \frac{(1/2)^2}{58} \right)} = 1.585.$$

The 95% confidence interval (using df = 140) is

$$L \pm t_{.025} SE_L$$

$$-1.15 \pm (1.977)(1.585)$$

$$-4.28 < \frac{1}{2}(\mu_{T,Hi} - \mu_{H,Hi}) + \frac{1}{2}(\mu_{T,Lo} - \mu_{H,Lo}) < 1.98 \text{ mm Hg.}$$

The confidence interval can also be written as

$$-4.28 < \mu_T - \mu_H < 1.98 \text{ mm Hg}$$

where

$$\mu_T = \frac{1}{2}(\mu_{T,Lo} + \mu_{T,Hi})$$

$$\mu_H = \frac{1}{2}(\mu_{H,Lo} + \mu_{H,Hi})$$

11.30 (b) Using the means given in Exercise 11.8, the value of L is

$$L = 9.81 - \frac{1}{2}(6.28 + 5.97) = 3.685 \text{ nmol/10}^8 \text{ platelets/hour.}$$

To find the standard error of L, we need to calculate s_{pooled}. From Exercise 11.7, SS(within) = 418,25 and df(within) = (18 + 16 + 8) - 3 = 39, so

$$s_{pooled} = \sqrt{\frac{418.25}{39}} = \sqrt{10.72} \, .$$

The multipliers in L are 1, $-\frac{1}{2}$, and $-\frac{1}{2}$. The standard error of L is

$$SE_L = \sqrt{10.72 \left(\frac{1}{18} + \frac{(-1/2)^2}{16} + \frac{(-1/2)^2}{8} \right)} = 1.048 \text{ nmol/10}^8 \text{ platelets/hour.}$$

11.33 The ordered array of means is

Treatment	C	A	B	E	D
Mean	3.70	4.37	4.76	5.38	5.41

We have $k = 5$ and $n = 9$. The scale factor is

$$\sqrt{\frac{MS(within)}{n}} = \sqrt{\frac{.2246}{9}} = .15797.$$

The total number of observations is $n^* = (5)(9) = 45$, so that $df(within) = 45 - 5 = 40$. We obtain values of q_j from Table 10 with $\alpha = .05$ and $df = 40$; we then multiply each q_j by .15797 to obtain R_j. The results are as follows:

j	2	3	4	5
q_j	2.86	3.44	3.79	4.04
R_j	.452	.543	.599	.638

The following table shows the sequence of comparisons:

j	Comparison					Conclusion
5	5.41 - 3.70	=	1.71	>	.638	Reject
4	5.41 - 4.37	=	1.04	>	.599	Reject
4	5.38 - 3.70	=	1.68	>	.599	Reject
3	5.41 - 4.76	=	.65	>	.543	Reject
3	5.38 - 4.37	=	1.01	>	.543	Reject
3	4.76 - 3.70	=	1.06	>	.543	Reject
2	5.41 - 5.38	=	.03	<	.452	Do not reject
2	5.38 - 4.76	=	.62	>	.452	Reject
2	4.76 - 4.37	=	.39	<	.452	Do not reject
2	4.37 - 3.70	=	.67	>	.452	Reject

The following hypotheses are rejected:

$H_0: \mu_C = \mu_D$	$H_0: \mu_A = \mu_E$
$H_0: \mu_A = \mu_D$	$H_0: \mu_C = \mu_B$
$H_0: \mu_C = \mu_E$	$H_0: \mu_C = \mu_A$
$H_0: \mu_B = \mu_D$	$H_0: \mu_B = \mu_E$

The following hypotheses are not rejected:

$H_0: \mu_A = \mu_B$ $H_0: \mu_E = \mu_D$

Summary:

C A B E D

There is sufficient evidence to conclude that treatments D and E give the largest means, treatments A and B the next largest, and treatment C the smallest. There is insufficient evidence to conclude that treatments A and B give different means or that treatments D and E give different means.

11.38 There are $_5C_2 = 10$ possible pairwise comparisons and we have df = 45 - 5 = 40. Thus, the Bonferroni multiplier is $t(40)_{.025/10}$, which we find in Table 11 to be 2.971. The Bonferroni method gives as the confidence interval

$$(5.38 - 4.37) \pm (2.971)\sqrt{.2246}\sqrt{\frac{1}{9} + \frac{1}{9}}$$

$1.01 \pm .664$
$(.346, 1.674)$ or $.346 < \mu_E - \mu_A < 1.674$.

11.42 The hypotheses are

H_0: The mean refractive error is the same in the four populations ($\mu_1 = \mu_2 = \mu_3 = \mu_4$)
H_A: Some of the populations have different mean refractive errors (the μ's are not all equal)

We have k = 4 and n* = 211. Thus,

df(between) = k - 1 = 4 - 1 = 3

df(within) = n* - k = 211 - 4 = 207

The ANOVA table is

Source	df	SS	MS
Between groups	3	129.49	43.16
Within groups	207	2506.8	12.11
Total	210	2636.3	

The test statistic is $F_s = \dfrac{MS(between)}{MS(within)} = \dfrac{43.26}{12.11} = 3.56$. With df = 3 and ∞, Table 9 gives $F_{.02} = 3.28$ and $F_{.01} = 3.78$. Thus, we have

$.01 < P < .02$.

Since $P < \alpha$, we reject H_0. There is sufficient evidence ($.01 < P < .02$) to conclude that some of the populations have different mean refractive errors.

11.44 k = 4; n = 3 for each group; n* = (4)(3) = 12.

df(between) = k - 1 = 3

df(within) = n* - k = 12 - 4 = 8

Source	df	SS	MS
Between groups	3	1.3538	.4513
Within groups	8	.27513	.03439
Total	11	1.6289	

The hypotheses are

H_0: The four treatments produce the same mean yield ($\mu_1 = \mu_2 = \mu_3 = \mu_4$)

H_A: Some of the treatments produce different mean yields (the μ's are not all equal)

The test statistic is $F_s = \dfrac{MS(\text{between})}{MS(\text{within})} = \dfrac{.4513}{.03439} = 13.1$. With df = 3 and 8, Table 9 gives $F_{.01} = 7.59$ and $F_{.001} = 15.83$. Thus, we have $.001 < P < .01$. Since $P < \alpha$, we reject H_0. There is sufficient evidence ($.001 < P < .01$) to conclude that some of the treatments produce different mean yields.

CHAPTER 12

Linear Regression and Correlation

12.1 (a) The slope and intercept of the regression line are

$$b_1 = \frac{\Sigma(x_i - \bar{x})(y_i - \bar{y})}{\Sigma(x_i - \bar{x})^2} = \frac{40}{10} = 4;$$

$$b_0 = \bar{y} - b_1\bar{x} = 13 - (4)(3) = 1.$$

The fitted regression line is $Y = 1 + 4X$.

For each point (x_i, y_i), the value of $\hat{y}_i$ is calculated as

$$\hat{y}_i = b_0 + b_1 x_i = 1 + 4x_i.$$

The values of x_i, y_i, and $\hat{y}_i$ are given in the following table:

x_i	y_i	$\hat{y}_i$
3	13	13
4	15	17
1	4	5
2	11	9
5	22	21

(c) $SS(resid) = \Sigma(y_i - \hat{y}_i)^2 = (13 - 13)^2 + (15 - 17)^2 + (4 - 5)^2 + (11 - 9)^2 + (22 - 21)^2 = 10.$

12.4 (b) The slope and intercept of the regression line are

$$b_1 = \frac{\Sigma(x_i - \bar{x})(y_i - \bar{y})}{\Sigma(x_i - \bar{x})^2} = \frac{6.92369}{.906191} = 7.64043 \approx 7.640;$$

$$b_0 = \bar{y} - b_1\bar{x} = 3.053 - (7.64043)(.4771) = -.592.$$

The fitted regression line is $Y = -.592 + 7.640X$.

$$s_{Y|X} = \sqrt{\frac{SS(resid)}{n - 2}} = \sqrt{\frac{10.8773}{13}} = .915°C.$$

12.7 (a) The slope and intercept of the regression line are

$$b_1 = \frac{\Sigma(x_i - \bar{x})(y_i - \bar{y})}{\Sigma(x_i - \bar{x})^2} = \frac{21953.7}{877.74} = 25.0116 \approx 25.01$$

$$b_0 = \bar{y} - b_1\bar{x} = 2168 - (25.0116)(62.40) = 607.3$$

The fitted regression line is $Y = 607.3 + 25.01X$.

(b)

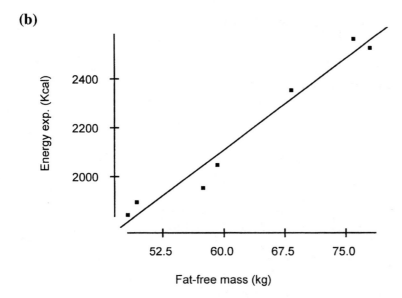

(c) As fat-free mass goes up by 1 kg, energy expenditure goes up by 25.01 Kcal, on average.

(d) $s_{Y|X} = \sqrt{\dfrac{SS(resid)}{n-2}} = \sqrt{\dfrac{21026.1}{5}} = 64.85$ Kcal

12.15 We begin by finding the equation of the regression line for the data in Exercise 12.6. The slope and intercept of the regression line are

$$b_1 = \frac{-927.75}{1303} = -.7120;$$

$$b_0 = \bar{y} - b_1\bar{x} = 23.64 - (-.7120)(11.5) = 31.83.$$

The fitted regression line is $Y = 31.83 - .7120X$.

Substituting $X = 15$ yields

$$Y = 31.83 - (.7120)(15) = 21.1.$$

Thus, we estimate that the mean fungus growth would be 21.1 mm at a laetisaric acid concentration of 15 µg/ml.

According to the linear model, the standard deviation of fungus growth does not depend on X. Our estimate of this standard deviation, $\sigma_{Y|X}$, is the residual standard deviation from the regression line, $s_{Y|X}$.

$$s_{Y|X} = \sqrt{\frac{SS(resid)}{n-2}} = \sqrt{\frac{16.7812}{12-2}} = 1.3 \text{ mm.}$$

Thus, we estimate that the standard deviation of fungus growth would be 1.3 mm at a laetisaric acid concentration of 15 µg/ml.

12.19 (a) We begin by calculating b_1 from the data in Exercise 12.3:

$$b_1 = \frac{81.90}{2800} = .02925.$$

To construct a 95% confidence interval, we consult Table 4 with $df = n - 2 = 7 - 2 = 5$; the critical value is $t(4)_{.025} = 2.571$. The confidence interval is

$$b_1 \pm t_{.025}SE_{b_1}$$

$$.02925 \pm (2.571)(.00159)$$

$$.0252 < \beta_1 < .0333 \text{ ng/min.}$$

(b) We are 95% confident that the rate at which leucine is incorporated into protein in the population of all *Xenopus* oocytes is between .0252 ng/min and .0333 ng/min.

12.23 From Exercise 12.7

$$b_1 = \frac{\Sigma(x_i - \bar{x})(y_i - \bar{y})}{\Sigma(x_i - \bar{x})^2} = \frac{21953.7}{877.74} = 25.0116 \approx 25.01;$$

$$s_{Y|X} = \sqrt{\frac{SS(resid)}{n - 2}} = \sqrt{\frac{21026.1}{5}} = 64.85.$$

Thus, the standard error of the slope is

$$SE_{b_1} = \frac{s_{Y|X}}{\sqrt{\Sigma(x_i - \bar{x})^2}} = \frac{64.85}{\sqrt{877.74}} = 2.189.$$

(a) To construct a 95% confidence interval we consult Table 4 with $df = n - 2 = 7 - 2 = 5$; the critical value is $t(4)_{.025} = 2.571$. The confidence interval is

$$b_1 \pm t_{.025}SE_{b_1}$$

$$25.01 \pm (2.571)(2.189)$$

$$19.4 < \beta_1 < 30.6 \text{ Kcal/kg.}$$

(b) To construct a 90% confidence interval we consult Table 4 with $df = n - 2 = 5$; the critical value is $t(4)_{.05} = 2.015$. The confidence interval is

$$b_1 \pm t_{.025}SE_{b_1}$$

$$25.01 \pm (2.015)(2.189)$$

$$20.6 < \beta_1 < 29.4 \text{ Kcal/kg.}$$

12.24 The hypotheses are

H_0: There is no linear relationship between respiration rate and altitude of origin ($\beta_1 = 0$)

H_A: Trees from higher altitudes tend to have higher respiration rates ($\beta_1 > 0$)

The sample slope was found in Exercise 12.8 to be

$$b_1 = \frac{161.40}{506667} = .0003186.$$

We note that $b_1 > 0$, so the data do deviate from H_0 in the direction specified by H_A.

From Exercise 12.8, the residual standard deviation is

$$s_{Y|X} = \sqrt{\frac{.013986}{10}} = .0374.$$

The standard error of the slope is

$$SE_{b_1} = \frac{s_{Y|X}}{\sqrt{\Sigma(x_i - \bar{x})^2}} = \frac{.0374}{\sqrt{506667}} = .00005254.$$

The test statistic is

$$t_s = \frac{b_1}{SE_{b_1}} = \frac{.0003186}{.00005254} = 6.06.$$

Consulting Table 4 with df $= n - 2 = 12 - 2 = 10$, we find that $t_{.0005} = 4.587$. Thus, $P < .0005$, so we reject H_0. There is sufficient evidence ($P < .0005$) to conclude that trees from higher altitudes tend to have higher respiration rates.

12.29 (a) $r = \dfrac{\Sigma(x_i - \bar{x})(y_i - \bar{y})}{\sqrt{\Sigma(x_i - \bar{x})^2 \Sigma(y_i - \bar{y})^2}} = \dfrac{21953.7}{\sqrt{(877.74)(570124)}} = .981.$

(b) The standard deviation of Y is

$$s_Y = \sqrt{\frac{\Sigma(y_i - \bar{y})^2}{n - 1}} = \sqrt{\frac{570124}{7 - 1}} = 308.25 \text{ Kcal.}$$

The residual standard deviation is

$$s_{Y|X} = \sqrt{\frac{SS(resid)}{n - 2}} = \sqrt{\frac{21026.1}{5}} = 64.85 \text{ Kcal.}$$

The approximate relationship is

$$\frac{s_{Y|X}}{s_Y} \approx \sqrt{1 - r^2}.$$

Substituting the values from above gives

$$\frac{s_{Y|X}}{s_Y} = \frac{64.85}{308.25} = .210.$$

From part (a)

$$\sqrt{1 - r^2} = \sqrt{1 - .981^2} = .194.$$

We verify that $.210 \approx .194$.

12.32 (a) $r = \dfrac{\Sigma(x_i - \bar{x})(y_i - \bar{y})}{\sqrt{\Sigma(x_i - \bar{x})^2 \Sigma(y_i - \bar{y})^2}} = \dfrac{893.689}{\sqrt{(1419.82)(853.396)}} = .812.$

12.34 The hypotheses are

H_0: There is no correlation between blood urea and uric acid concentration in the population of all healthy persons ($\rho = 0$)

H_A: Blood urea and uric acid concentration are positively correlated ($\rho > 0$)

The test statistic is

$$t_s = r\sqrt{\frac{n-2}{1-r^2}} = .2291\sqrt{\frac{282}{1-.2291^2}} = 3.952.$$

The degrees of freedom are 284 - 2 = 282, so we consult Table 4 using df = 140. We find that $t(140)_{.0005} = 3.361$. Thus, P < .0005, so we reject H_0. There is sufficient evidence (P < .0005) to conclude that blood urea and uric acid concentration are positively correlated.

12.44 Let X = body weight and let Y = ovary weight. We wish to estimate $\sigma_{Y|X=4}$. According to the linear model, $\sigma_{Y|X}$ does not depend on X; thus, our estimate of $\sigma_{Y|X}$ will be $s_{Y|X}$. According to Fact 12.1,

$$\frac{s_{Y|X}}{s_Y} \approx \sqrt{1-r^2}.$$

We are given that $s_Y = .429$ gm and r = .836. Thus, we have

$$\frac{s_{Y|X}}{.429} \approx \sqrt{1-.836^2}.$$

Solving this equation yields

$$s_{Y|X} \approx (.429)\sqrt{1-.836^2} = .24.$$

Hence, we estimate $\sigma_{Y|X=4}$ to be .24 gm.

12.46 (a) We begin by calculating the fitted regression line for the data in Exercise 12.37. The slope and intercept of the regression line are

$$b_1 = \frac{\Sigma(x_i - \bar{x})(y_i - \bar{y})}{\Sigma(x_i - \bar{x})^2} = \frac{-.342}{.1512} = -2.262;$$

$$b_0 = \bar{y} - b_1\bar{x} = 1.117 - (-2.262)(.12) = 1.39.$$

The fitted regression line is Y = 1.39 - 2.262X. Substituting X = .24 yields

$$Y = 1.39 - (2.262)(.24) = .85.$$

Thus, we estimate that the mean yield would be .85 kg at a sulfur dioxide concentration of .24 ppm.

According to the linear model, $\sigma_{Y|X}$ does not depend on X; thus, our estimate of $\sigma_{Y|X}$ will be $s_{Y|X}$. We are given that SS(resid) = .2955. The residual standard deviation is

$$s_{Y|X} = \sqrt{SS(resid)/(n-2)} = \sqrt{.2955/10} = .17 \text{ kg.}$$

Thus, we estimate that the standard deviation of yields would be .17 kg at a sulfur dioxide concentration of .24 ppm.

12.49 (a) $r = \dfrac{\Sigma(x_i - \bar{x})(y_i - \bar{y})}{\sqrt{\Sigma(x_i - \bar{x})^2 \Sigma(y_i - \bar{y})^2}} = \dfrac{2.01571}{\sqrt{(23.7143)(.265486)}} = .803.$

(b) The standard deviation of Y is

$$s_Y = \sqrt{\frac{\Sigma(y_i - \bar{y})^2}{n-1}} = \sqrt{\frac{.265486}{7-1}} = .210 \text{ cm.}$$

The residual standard deviation is

$$s_{Y|X} = \sqrt{\frac{SS(resid)}{n-2}} = \sqrt{\frac{.09415}{5}} = .137 \text{ cm.}$$

The approximate relationship is

$$\frac{s_{Y|X}}{s_Y} \approx \sqrt{1 - r^2}.$$

Substituting the values from above gives

$$\frac{s_{Y|X}}{s_Y} = \frac{.137}{.210} = .652.$$

From part (a)

$$\sqrt{1 - r^2} = \sqrt{1 - .803^2} = .596.$$

We verify that $.652 \approx .596$.

CHAPTER 13

A Summary of Inference Methods

13.1 The response variable in the study is whether or not a patient shows clinically important improvement; this is a categorical variable. The predictor variable is group membership: clozapine or haloperidol. The two samples are independent and the sample sizes are rather large. Thus, a chi-square test of independence would be appropriate. The null hypothesis of interest is H_0: $p_1 = p_2$, where

$p_1 = \Pr\{$clinically important improvement if given clozapine$\}$

and

$p_2 = \Pr\{$clinically important improvement if given haloperidol$\}$.

A confidence interval for $p_1 - p_2$ would also be relevant.

13.9 A two-sample comparison is called for here, but the normal probabilty plots shown below indicate that the data do not support the condition of normality. Thus, the Wilcoxon-Mann-Whitney test is appropriate.

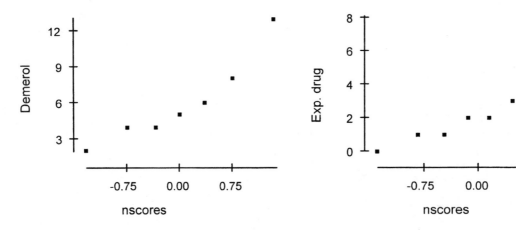

13.11 It would be natural to consider correlation and regression with these data. For example, we could regress Y = forearm length on X = height; we could also find the correlation between forearm length and height and test the null hypothesis that the population correlation is zero.